"*Power in Conservation* is that wide-ranging and accessible but cutting-edge introduction to contemporary conservation, social science and theory for which students, researchers, and practitioners alike have been searching. Carpenter provides to the curious reader a deep analytic exploration, one that will repay reading with useful tools and unexpected insights in every chapter of this brilliant book."

Arun Agrawal, Professor, School for Environment and Sustainability, University of Michigan, USA

"In this bold, workmanlike and approachable overview of the anthropological literature, Carol Carpenter engages with the theoreticians and significant ethnographic studies that underpin our emerging understanding of a new Anthropocene-shaped world. It is Foucault and his interpreters who dominate this new synthesis, and nuanced ethnographic accounts that show us the links between power and knowledge, sovereignty and governmentality, and which take us beyond the paradigm of 'political ecology'. A magnificent synthesis."

Roy Ellen, Centre of Biocultural Diversity, University of Kent, UK

"Carol Carpenter provides a clear and usable—but not simplified—theoretical tool kit for understanding power in conservation, and explains why a fine-grained, ethnographic attention to the workings of power is essential for effective conservation practice. Her rigor is impressive, and the implications of her analysis are profound. A masterful contribution."

Tania Li, Professor of Anthropology, University of Toronto, Canada

POWER IN CONSERVATION

This book examines theories and ethnographies related to the anthropology of power in conservation.

Conservation thought and practice is power laden—conservation thought is powerfully shaped by the history of ideas of nature and its relation to people, and conservation interventions govern and affect peoples and ecologies. This book argues that being able to think deeply, particularly about power, improves conservation policy-making and practice. Political ecology is by far the most well-known and well-published approach to thinking about power in conservation. This book analyzes the relatively neglected but robust anthropology of conservation literature on politics and power outside political ecology, especially literature rooted in Foucault. It is intended to make four of Foucault's concepts of power accessible, concepts that are most used in the anthropology of conservation: the power of discourses, discipline and governmentality, subject formation, and neoliberal governmentality. The important ethnographic literature that these concepts have stimulated is also examined. Together, theory and ethnography underpin our emerging understanding of a new, Anthropocene-shaped world.

This book will be of great interest to students and scholars of conservation, environmental anthropology, and political ecology, as well as conservation practitioners and policy-makers.

Carol Carpenter is Senior Lecturer in the School of Forestry and Environmental Studies at Yale University, US. She is co-editor of *Environmental Anthropology: A Historical Reader* (2007).

Routledge Studies in Conservation and the Environment

This series includes a wide range of inter-disciplinary approaches to conservation and the environment, integrating perspectives from both social and natural sciences. Topics include, but are not limited to, development, environmental policy and politics, ecosystem change, natural resources (including land, water, oceans and forests), security, wildlife, protected areas, tourism, human-wildlife conflict, agriculture, economics, law and climate change.

Natural Resources, Tourism and Community Livelihoods in Southern Africa
Challenges of Sustainable Development
Edited by Moren T. Stone, Monkgogi Lenao and Naomi Moswete

Leaving Space for Nature
The Critical Role of Area-Based Conservation
Nigel Dudley and Sue Stolton

Power in Conservation
Environmental Anthropology Beyond Political Ecology
Carol Carpenter

For more information about this series, please visit: www.routledge.com/Routledge-Studies-in-Conservation-and-the-Environment/book-series/RSICE

POWER IN CONSERVATION

Environmental Anthropology Beyond Political Ecology

Carol Carpenter

Routledge
Taylor & Francis Group
LONDON AND NEW YORK

earthscan
from Routledge

First published 2020
by Routledge
2 Park Square, Milton Park, Abingdon, Oxon OX14 4RN

and by Routledge
52 Vanderbilt Avenue, New York, NY 10017

Routledge is an imprint of the Taylor & Francis Group, an informa business

British Library Cataloguing-in-Publication Data
A catalogue record for this book is available from the British Library

Library of Congress Cataloging-in-Publication Data
A catalog record has been requested for this book

ISBN: 978-0-367-34251-7 (hbk)
ISBN: 978-0-367-34250-0 (pbk)
ISBN: 978-0-429-32465-9 (ebk)

Typeset in Bembo
by Wearset Ltd, Boldon, Tyne and Wear

CONTENTS

ACKNOWLEDGMENTS

I would like to thank the authors of books that have inspired me, particularly ethnographies. I begin with the three books that changed everything for a Ph.D. anthropologist flailing around in the development world: James Ferguson's *The Anti-Politics Machine*, Arturo Escobar's *Encountering Development*, and James Fairhead and Melissa Leach's *Misreading the African Landscape*. Once I bailed out of development contracting, I began to teach, and discovered what was at the end of the twentieth century a fairly small pool of publications. Most of the readings I found to teach initially concerned development, but the literature on conservation soon caught up. I read and taught Tania M. Li (who now has two major ethnographies, *Will to Improve* and *Land's End*), Anna L. Tsing (who also now has two major works, *Friction* and *Mushroom at the End of the World*), Donald Moore (*Suffering for Territory*), William Cronon (*Uncommon Ground* and *Nature's Metropolis*), Arun Agrawal (*Environmentality*), Timothy Mitchell (*Rule of Experts*), Bruce Braun (*The Intemperate Rainforest*), David Mosse (*Cultivating Development*), and Stephen Gudeman (especially *Conversations in Colombia*). Li, Moore, and Gudeman have joined my classes, and I have heard Escobar, Li, Tsing, Moore, and Cronon give talks. But mostly I thank all of you, and many more I haven't mentioned, for your published words.

This book owes a tremendous debt to Tania Li's writing. Li has produced a corpus: decades of thinking about multiple sites in Sulawesi, Indonesia, educating herself and others in the theory needed to best understand them.

I also want to thank my students. Being a lecturer at Yale has meant that I teach a lot, but it has also meant that I have outstanding students, who mostly share my excitement about the literature above, and in this book as a whole. In addition, I need to thank two particular classes. My "Social Science of Conservation and Development" class Fall 2017 endured the first draft of this book in the form of lectures, and produced outstanding final papers. They include: my Teaching Fellow, Cara Donovan; Lottie Boardman, Vivian Breckenridge, Frank Cervo, Paula Chamas Piedrabuena, Corey Creedon, Blair Crossman, Gyan de Silva, Yufang Gao, Javier Gonzalez, Cori Grainger, Greg Haber, Thomas Launer, Zander Pellegrino, Elham Shabahat, and Alberto Tordesillas Torres. Fall 2018 students in the same course were able to read second drafts of the book, which gave us more time to talk about the literature. Those discussions were excellent, and incredibly helpful to me. And again, the students wrote outstanding papers. They include: my Teaching Fellow, Lottie Boardman; Ashia Ajani, Sofia Caycedo, Britta Dosch, Vanessa Koh, Nora Moraga-Lewy, Jorge Nieto Jimenez, Kaggie Orrick, Alix Pauchet, MK Speth, and Brittany Wienke. You all made a contribution to this book, and I thank you for listening to me and reading my words.

I give my heartfelt thanks to my Dean, Indy Burke, for finding some time for me to work on this book. Without that it would literally not have been written.

My editor at Routledge, Hannah Ferguson, and assistant editor John Baddeley, have been supportive throughout.

I want to thank my sisters, Diane Carpenter and Janet Carpenter Sotola, who supported me throughout this project. And I thank my daughter, Margaret Carpenter-Dove, who has been excited about this book since I first dreamed of doing it.

1

INTRODUCTION

> I would like my books to be a kind of tool-box which others can rummage through to find a tool which they can use however they wish in their own area ... I don't write for an audience, I write for users, not readers.
>
> (Foucault 1994:523–4)

Introduction

Just as Foucault intended to provide a "tool-box" for users not readers, this book is intended to give the reader a tool-box of ideas about power in order to improve the effectiveness of conservation, particularly in the global South. Conservation thought and practice is power laden. Interventions govern and affect peoples and also ecologies, especially in the global South. I think it is important to accept that these interventions are typically well meaning towards people as well as valuing the environment. But conservation interventions are still full of power—and not just a single sort of power. I argue that being able to think deeply and specifically, particularly about power, improves conservation policy-making and practice.

Most of the power in conservation doesn't lie with money and influence, nor with coercive control, though there is some of that. A lot has been written about those sorts of power. But most of the power in conservation lies with mistaken assumptions we hold about people, misunderstandings about their relation with the environment and with the economy, and assumptions about our own superior knowledge, and our sense that we have a right to intervene based on that assumed superior knowledge. Those interventions cannot help but constitute some sort of governing. Most power is well meaning. Most power in conservation doesn't even know itself as power or as governing. This book is about those sorts of power.

They are much more subtle and hard to understand than money and coercion, but they are pervasive in conservation. We need to open our eyes to the prevalence of power in conservation.

Before we begin to examine ideas of power, however, it is essential that we think about the contemporary context in which conservation practices. We also need to face the lack of contemporary success in conservation, and the dearth of data evaluating conservation.

Conservation

According to the Red List published by the IUCN (International Union for Conservation of Nature, n.d.), more than 28,000 species, 27 percent of all assessed species, are threatened with extinction. According to BirdLife International's *State of the World's Birds*, at least 40 percent of the world's birds have declining populations; this is one in eight of all birds (2018, 11). Forests are also decreasing: according to the World Bank, forest area (and forest area as percentage of land area) has decreased about one million square kilometers between 1990 and 2016—a time period that includes interventions to protect forests around the world (The World Bank Data 2019). There is a very real need for conservation interventions to be effective.

The history of such interventions goes back about 70 years,[1] but we know little about their effectiveness, and what we do know suggests that they are failing. In 2019 a meeting of 70 leading natural and social science academics, professionals, and researchers was held in Vienna to discuss the reasons for the failure of conservation (Vidal 2019). A 2013 publication in *Conservation Biology* (by Fisher et al.) is entitled "Moving Rio Forward and Avoiding 10 More Years with Little Evidence for Effective Conservation Policy." A 2016 review by Baylis et al. (58) concludes that "despite a critical need for empirical evidence, conservation science has been slow to adopt [modern] impact evaluation designs." Evidence of the effect of protected areas, payment for ecosystem services (PES), and forest management on human welfare, habitats, and species preservation is "limited or fragmented"; for things like forest law enforcement, PES schemes, ecocertification, and integrated conservation and development approaches, evidence "is even more sparse" (Baylis et al. 2016: 61). The best evidence concerns terrestrial protected areas, but even there it is limited to habitat loss rather than species preservation, and is primarily limited to forests (especially tropical ones), which can be measured with remote sensing (Geldmann et al. 2013). Moreover, "The effect size of protection is small": protected forests usually fare better than non-protected ones (often in buffer zones), but if threats are high protected habitat loss can also be high (Geldmann et al. 2013). A 2016 article similarly concludes that the nine studies it reviewed "estimated that annual conservation impacts on forest cover were below one percent, with two exceptions" (Börner et al. 2016). In short, the effectiveness of conservation is not being adequately evaluated, and even the most adequately evaluated sector, protected areas, seems to have low levels of effectiveness.

When the effects of conservation on people are considered, data collection is even more challenged. Since 1992 in Rio, conservation is routinely paired with development. What this has meant is that many conservation projects have social goals also, particularly poverty alleviation (the idea being that poverty leads to overuse of the environment). An example that predates Rio is community forest management: a 2012 review found evidence of greater tree density, but "no data on local human welfare amenable to meta-analysis" (Bowler et al. 2012:29); a 2015 review simply says that "many projects have failed, either partly or completely" (Baynes et al. 2015:226). Payments for environmental services tells a similar story: "Progam effectiveness often lags behind the expectations of early theorists" (Börner et al. 2017).

Baynes et al. (2015: 232–3) discern four themes in the failures of community forestry: first, "Power is rarely voluntarily shared" even within communities; second, many governments pay lip-service to community forestry but reduce budgets, because "political opposition to genuine devolution of decision-making [from forestry departments] to CFGs [community forest groups] is often entrenched"; third, long-term material benefits are insufficient to motivate CFG members, who also need short-term cash income; finally, "Tenure security is the factor which most affects CFG members' motivation." The final factor involves all three of the other factors: insecure tenure argues against waiting for trees to reach harvest age; histories of tenure insecurity arising from the state go back to the colonial period (and, in my experience, are remembered); and power differentials pit users of major and minor forest products against each other. The bottom line is that tenure is insecure because of the power of governments over it, and that power differences within the community make even use rights insecure. Power limits the success of the community forestry model to attain both its conservation and its social goals.[2]

I argue that conservation alone, without development or even social justice goals, requires an understanding of both ecological and social aspects—particularly those linked to power. This is not a new argument. In 2003, in an Editorial to *Conservation Biology*, Mascia et al. (a team including both environmental and social scientists) argue:

> Although it may seem counterintuitive that the foremost influences on the success of *environmental* policy could be *social*, conservation interventions are the product of *human* decision-making processes and require changes in *human* behavior to succeed. Thus, conservation policies and practices are inherently social phenomena.
>
> *(649, authors' italics)*

Guerrero et al. (2018:37) similarly argue that "people and nature are inextricably linked. Overcoming sustainability challenges thus requires an integrated social-ecological science." Real integration is not unidirectional. Promoting the recovery of endangered species, for example:

> requires understanding the human activities affecting the viability of the species and also how species dynamics affect human livelihoods. Addressing sustainability challenges requires a shift in focus: from seeing people and nature as separate systems to seeing them as two interacting components of a complex, dynamic, and integrated system.
>
> *(Guerrero et al. 2018:37)*

Guerrero et al. (2018) found that true integration was still lacking in social-ecological research. They also found that certain social variables important for understanding challenges were "underrepresented": "culture, politics, and power" (Guerrero et al. 2018:37).

In sum, effective conservation is desperately needed, and it requires an integrated understanding of the ecological and the social—particularly, I argue, of politics and power.

Theories of power

Within this hypothetical tool-box, however, there is a division. There are two theoretical bases or roots to thinking about power in conservation: one is based on Marx, and the other on Foucault. In academic circles, the two are sometimes opposed to each other, and sometimes used side by side without recognition of the division—especially in political ecology since the mid-1990s. In order to think effectively about conservation, we need to be able to use both Marxist and Foucauldian ideas about power, and we need to use both well. But we should begin by thinking more deeply about the difference between the two.

Marxist and Foucauldian approaches to power differ most essentially in the way they view the relation between the economy and politics or power. Marx separates economy,[3] as base, from power, as political superstructure,[4] and base to some extent determines superstructure. The base–superstructure relation in Marx is actually more complex than the idea that an economic base determines a political superstructure, though as we will see this relation is often over-simplified in later Marxist thought. In Marx's writing there is some reciprocity between base and superstructure; Engels made this explicit, writing that the relation was only determined "in the last instance" (Calhoun 2002:34). Foucault goes farther. He does not separate economy and power.

Educated in Marxist academic France, Foucault was in fact reacting against Marxist views of power. He argued that power does not flow from the economy (1980:88, 116). He asks: "Is the analysis of power or of powers to be deduced in one way or another from the economy?" and "Is power always in a subordinate position relative to the economy?"

> Or, on the contrary, do we need to employ varying tools in its analysis—even, that is, when we allow that it effectively remains the case that the

> relations of power do indeed remain profoundly enmeshed in and with economic relations and participate with them in a common circuit?
>
> *(1980:88, 89)*

In Foucault, relations of power are "profoundly enmeshed in and with" economic relations; they participate in a "common circuit." Power relations are not in a position of exteriority to economic processes, but are immanent in them (Foucault 1978:94). He also says that the new form of power of the seventeenth and eighteenth centuries (which will become governmentality) concerns itself with the accumulation of people, the population, and is "correlated and inseparable" from the new economic system, capitalism, which promotes the accumulation of capital (Foucault 1980:125, 133). In order to add Marxist and Foucauldian understandings to each other, we need to question, not assume, the role of economic forces.

Marx and Engels' concept of the base–superstructure relation lent itself to over-simplification in time, as Marxist approaches tended to focus on the economic. The cultural Marxist Raymond Williams, in a 1973 essay, explicitly criticizes this.[5] As he writes, "In the transition from Marx to Marxism, and in the development of mainstream Marxism itself, the proposition of the determining base and the determined superstructure has been commonly held to be the key to Marxist cultural analysis" (Williams 2005:31). Williams revalues all three of the key terms. "Determination" becomes "the setting of limits and the exertion of pressure"; "superstructure" becomes "a related range of cultural practices" (to translate this from culture to power, we could suggest "a related range of political practices"); and base becomes not a fixed abstraction, but "the specific activities of men in real social and economic relationships, containing fundamental contradictions and variations and therefore always in a state of dynamic process" (Williams 2005:34).

In a 2010 book review forum on *The Will to Improve*, Tania Li writes that one of her objectives in the book was "to complicate a critical narrative that connects development interventions too quickly, and too directly, to capitalist accumulation" (Silvey 2010:233). Conservation, for example, contributed to land dispossession by closing off the gradual expansion of agriculture into the forest that had prevented dispossession for generations—but was "disconnected from accumulation, as conservation is seldom profitable" (Silvey 2010:233). Her analytical choice was "to examine the mechanisms of dispossession in all their particularity, avoiding the implicit functionalism of some linear accounts" (Silvey 2010:233). Though she does not name political ecology, I believe it is the critical narrative she complicates.

Writing in 1996, the anthropologist Donald S. Moore (126) says that "much political ecology remains within a macro-structural framework that emphasizes the determining influence of broad economic forces." He argues rather that micropolitics (peasant struggles over access to natural resources), symbolic practices, and structural forces mutually construct one another. I find that Foucault's ideas about power are also over-simplified, particularly when they are adopted by political ecologists, as I will argue below.

In sum, I argue that only a tool-box of ideas about power will improve conservation. The biggest impediment to constructing such a tool-box is the difference between Marxist and Foucauldian approaches to the relation between the economy and power. This difference is exaggerated by political ecologists who over-simplify both Marxist and Foucauldian ideas of power.

Political ecology

Political ecology is the most well-known approach to thinking about power in conservation. A great deal has been written about this approach, including at least five books (Perreault, Bridge, and McCarthy 2019; Bryant 2015; Robbins 2012; Biersack and Greenberg 2006; Zimmerer and Bassett 2003). Yet the way that political ecologists think about power, following Moore, tends to be Marxist. In Peet and Watts' (2004:8) second edition of *Liberation Ecologies*, the authors note the importance for the emergence of political ecology of "the growth of Marxism in the social sciences, especially in development studies in a variety of guises (world systems theory, dependency, structural Marxism) during the late 1960s and 1970s." Political ecology tends to be theorized as ecology plus political economy, and in particular Marxist political economy. As Paul Robbins' (2012:59, author's italics) text explains: "Almost all research in political ecology is theoretically engaged with what has often been described as a *broadly defined political economy.*" Here he is following (and cites) definitions of the field by Peet and Watts (1996), Greenberg and Park (1994:1), and Blaikie and Brookfield (1987) (Robbins 2012:15).[6]

The over-simplification of Marx's thought about the base–superstructure relation is clearly discernible in political ecology: politics tends to boil down to economics. As Robbins (2012:59) explains, "While not all of contemporary political ecology is explicit in its allegiance to materialism, much of the work at least tacitly assumes many materialist precepts." The first such precept is that "social and cultural relationships are rooted in economic interactions amongst people and between people and non-human objects and systems" (Robbins 2012:59).

Exaggerating the determining power of the economic base is clearly at odds with Foucault's understanding of the economy. In some political ecology, it leads to other, related simplifications of power. As Robbins[7] (2012:59) continues, concluding a section entitled "Marxist Political Economy":

> The systems that govern use, overuse, degradation, and recovery of the environment are structured into a larger social engine, which revolves around the control of nature and labor.... No explanation of environmental change is complete, therefore, without serious attention to who profits from changes in control over resources, and without exploring who takes what from whom.

The "larger social engine" and the setting up of an opposition between those who profit and extract and their victims, winners and losers, are typical of much

Marxism as well as much political ecology. And they are all at odds with Foucault's approach to power.[8]

The idea of some sort of "larger social engine" or "structural forces" characterizes much political ecology. There is not a single term for this; Robbins (2012:54) uses "political economic forces" and "fundamental economic forces." In response to Vayda and Walters' 1999 and Vayda's 2009 critiques of political ecology, Robbins (2012:150) says that they distrust "higher-order categories or emergent properties of relations," because "these reified categories often come to stand in for … direct and meaningful influences, in explanations where 'forces' become independent agents." Robbins (2012:153, 150) partially agrees with this, saying that political ecology "would benefit from decreased reification of conceptual categories," but he is also troubled by this critique, because it "refutes many 'structural' explanations," he writes, and "contradicts some key elements of political economy, which rely on well-established structural variables."

And what are these structural variables? Robbins' response to Bruno Latour is revealing. Latour, according to Robbins (2012:77), critiques "theoretical baggage like 'structures' or 'systems' that might otherwise mystify the concrete relationship between things." Against this "apolitical" critique, political ecology is "rooted in the notion that persistent outcomes are caused, at least in part, by structural constraints imposed by existing logics and processes in the world like *capitalism, racism, colonialism, and sexism*" (Robbins 2012:77, my italics). Elsewhere, Robbins (2012:13, 19, my italics) uses the terms "linked axes of *money, influence, and control*," "part of systems of *power and influence*." Note that Donald Moore, while critiquing determination by such forces, retains the forces themselves (co-constructed with micropolitics and symbolic practices). I would characterize the "structural forces" that appear in political ecology as the economic base—which Foucault directly questions. To sum up, political ecology is rooted in the idea that persistent outcomes, including overuse and degradation of the environment, are caused—at least in part—by political-economic forces which link money to power (influence and control), and are expressed in historical processes of inequality like capitalism, racism, colonialism, and sexism.

The opposition between those who profit and their victims is also typical of political ecology. Earlier in his text, Robbins (2012:87) says that there are four characteristics of political ecologies as texts, and the first is that they "track winners and losers to understand the persistent structures of winning and losing." Tracking winners and losers sets up a conceptual opposition that tends to map onto elites and peasants or indigenous people, for example—reproducing Marx's class relation in other terms. In Robbins' (2012:181) chapter on conservation and control, he suggests that conservation has failed because its instruments "have disenfranchised traditional land managers and enforced the goals, desires, and benefits of elite communities." Here the victims are the land and its traditional managers, and elite communities (which could be local elites, colonial elites, or international elites) have benefitted. If political ecologists track class relations to understand persistent structures, they are going to find some version of capitalism.

Class under capitalism thus seems to explain the failure of conservation. As Arun Agrawal (2005b:215) says, political ecologists view the environment as an arena in which oppositional conflicts unfold, between, for example, elite and poor, state and community, outsider and local. This constitutes a "constraint on the analytical imagination." "The argument often turns into the assignment of credit and blame." (Agrawal 2005b:215)

In political ecology, tracking winners and losers at the scale of communities and landscapes should (as first laid out by Blaikie and Brookfield (1987)) constitute the bottom link in a chain of explanation, which ultimately ascends to the state and the global economy (Robbins 2012:88–9). I would suggest that it is a quest for explanation that makes larger structures theoretically necessary: in political ecology, explanation depends on those structures or forces that are already presaged by the exercise of dividing people into winners and losers. And if larger economic structures tend to explain things in political ecology, then these structures are taking on the old, over-simplified role of determination.

Foucault avoids this by giving up on explanation: in his thought the question "why"—the quest for explanation—is supplanted by the question "how." Foucault asks how, not why, questions: "To begin the analysis with a 'how' is to suggest that power as such does not exist." He means "how" in the sense of "by what means is it exercised" (Foucault 1983:216–17). He is suggesting that how questions allow a deeper questioning of power, whereas why questions reify power as something prior to its exercise. Arun Agrawal (2005b:210, my italics) has a similar critique of political ecology: "The primacy accorded the political often prevents political-ecological analysis from examining *how* the political itself is made." It "casts politics as the prime mover, the cause that exists sui generis. But the *exercise* of power and political asymmetries … are themselves a consequence of many different processes" (Agrawal 2005b:210, my italics).

I argue that the quest for an explanation that ascends to large-scale structural forces causes an over-simplification of more complex concepts of power, like those of Foucault. Robbins is noteworthy for including several of Foucault's concepts in political ecology, but they are inflected by his understanding of power.[9] Presenting Foucault's concept of power/knowledge, for example, Robbins says that those things accepted as "truth" "*enforce* social order" and maintain "the power of individuals or groups," that they are "*conditioned and stabilized* by social structures of power" and "*formed and made powerful* by state institutions, media companies, experts, and families" (2012:70, my italics). This is an understanding of "truth" in which already powerful institutions create and lend power to "truth" in order to maintain their own power and the social order in which they are elites. The economy plays a more hidden role here, but may be lurking behind terms like "social structures of power" and "the social order." This is not Foucault's "truth." In Foucault, though his concept of power/knowledge develops in time (see above), power is internal to knowledge; power/knowledge is not "formed and made powerful" by anyone. In addition, though powerful knowledges can produce and reinforce power, they simultaneously produce and reinforce resistance; they do not "enforce social order" or maintain anyone's power, since they are available to both sides of all power relations.

Agrawal (2005a:165–6) writes that "many of those who extend Foucault's ideas about governmentality to the colonial and postcolonial contexts" remain "preoccupied mostly with the coercive aspects of state, institutional, and social power." Agrawal (2005b:216, 224) also describes that he uses Foucault's ideas as "a source of possible provocations" because:

> instead of taking power as the fixed property of some agent(s), it becomes possible to examine concretely how power is generated by and located in different strategies of government. No particular agent or person can then be seen as being located in a permanently more powerful political position vis-à-vis another.[10]

In sum:

> A focus on technologies of government and their application helps undermine the tendency to view institutions, power, or subject locations as the unquestionable starting point from which to gain an understanding of environmental change and politics. Instead it encourages attention to the processes through which these concepts are consolidated and naturalized. It directs analysis to the interdependent constitution of these three seemingly foundational concepts in environmental studies [institutions, power, and subject locations] and thereby makes the familiar contingent.
>
> *(Agrawal 2005b:230)*

Robbins over-simplifies governmentality in the same way. He elides governmentality with coercion and the internalization of state rule. It extends "beyond simply enforcing conservation rules, however. Rather, efforts center on extending the discretionary conservation power of the state by causing individuals and social groups to 'internalize' the coercive missions of the government, creating *self-enforcing* coercion" (179, author's italics).

In Foucault, governmentality must be understood as the opposite of coercion, since, like all power in Foucault except sovereignty, it is based on freedom (see Chapter 5). Governmental subject creation works by inducing individuals to observe, judge, and correct their own behavior, thereby conducting themselves towards some new purpose—but this new purpose is not state rule (see Chapter 9). Governmental subjects seek out experts defined by the human sciences, like psychiatrists or doctors, to try to improve themselves. Robbins may be thinking of disciplinary subject formation. The best example of this is the panopticon prison, in which the prisoners essentially internalize the guard because they cannot tell if he is watching them, but he could be (see Chapter 9). But even disciplinary subject formation can be resisted, and is more appropriately described as self-government than self-enforcing coercion.

In sum, I argue that the over-simplification of the base–superstructure relation, the explanatory role of economic "structural forces," the class relation as winners and losers, and the over-simplification of power as coercion originate in Marxist understandings, and lead to misunderstandings of Foucault's ideas about power. Power is difficult to understand. Foucault himself said, "I'm struck by the difficulty I had in formulating [power]" (1980:115). This book hopes to help make these ideas more accessible.

Foucault

The remainder of the book covers the anthropology of conservation literature on politics and power primarily from outside political ecology, especially literature rooted in Foucault. This literature is large and important. Foucault has had a seminal influence on the anthropology of conservation. This book is intended to illuminate, to make useful, four of Foucault's concepts of power that are most used in the anthropology of conservation:

1 *The power of discourses.* Conservation itself is a discourse with a great deal of power over our thinking. It carries deep assumptions about a dichotomy between the environment and people, for example, which are difficult to question. This concept is probably the most popular and used of Foucault's ideas, but it is also often over-simplified. Chapters 2, 3, and 4 explain Foucault's concept and explore it in ethnographies.
2 *The triangle.* Three different models of or ways of thinking about governing—sovereignty, discipline, and governmentality—often work alongside each other to shape conservation interventions. Governmentality is frequently used to describe conservation policy and practice; I argue that we should not forget the whole triangle. I also argue that discipline and governmentality are quite distinct ideas. Sovereignty, discipline, and governmentality constitute three additional tools for our tool-box of ideas of power in conservation. Chapters 5 and 6 explain Foucault's ideas and examine ethnographic examples of them.
3 *Subject formation.* Conservation interventions often attempt to make local peoples into conservationists; this is subject formation. Subjects are formed by discourses, and discipline and governmentality cultivate subjects quite differently. Chapters 9 and 10 examine this idea and its use in ethnographies.
4 *Neoliberal governmentality.* Conservation is increasingly governed through the market, by offering market opportunities in exchange for losses to conservation, including shares in the carbon market. This is neoliberal governmentality. In addition, conservation interventions may demand local financial contributions or local financial sustainability and/or teach cost–benefit analysis in trainings. This is neoliberal subject formation with governmentality.

Chapters 11 and 12 explain Foucault's concept and explore ethnographic examples of it.

Michel Foucault was born in 1926 in France. His young adulthood was colored by his homosexuality, which led him to leave France—at that time "an old traditional society," in which "freedom for personal life was very sharply restricted"—for Sweden—which was "supposed to be a much freer country"—in 1955 (Bernauer and Keenan 1987:271). In Sweden he learned that "a certain kind of freedom may have, not exactly the same effects, but as many restrictive effects as a directly restrictive society," which was "an important experience for me" (Bernauer and Keenan 1987:271). We can see this as the root of Foucault's understanding of a liberal power, based on freedom

Foucault completed his doctorate in philosophy in 1961, from the Ecole Normale Superieure. His dissertation became *Madness and Civilization* (published in English in 1965), which "failed to interest anyone" (Foucault 1980:110). He became a professor of philosophy in 1962. But his work didn't become acceptable until the May 1968 strikes in France, which were exceptionally non-Marxist. They created a "time of unexpected political opening," in which the "concrete nature of power, its techniques and tactics, became visible," making his past work acceptable and his future work possible, as the centrality of Marxism in academic France declined (Foucault 1980:111, 116). He was elected to the College de France in 1969, naming his chair "History of Systems of Thought." Foucault died of AIDS in 1984, at the age of 58 (Bernauer and Keenan 1987:273, 276).

Foucault worked very hard to think about difficult-to-think things. This means that his thinking evolved over time, so it's important to keep in mind when he actually wrote something or delivered a lecture, not the English-language publication date. The overarching themes of his work evolved.[11] His first theme was knowledge, specifically the discursive practices that articulated the human sciences.[12] In these early works power plays no role: in 1980 Foucault (113) says that in these works he was confusing discursive *regimes* with something more structural, like a paradigm, a theory, or the systematicity of knowledge.

By 1975, and probably beginning in 1968, his second theme—really a revelation—was power: the power inherent in scientific discourses (1980:109–10). This theme is evident in the 1975 (1977 in English) *Discipline and Punish*, and of course the 1980 *Power/Knowledge*, an English-language collection of writings and interviews dated between 1972 and 1977. *Discipline and Punish* emphasized practices and technologies, while *Power/Knowledge* emphasized knowledge or discourses. Both of these books were enormously influential in the social sciences in the United States, including environmental anthropology and political ecology. This theme is also clear in the 1976 Introduction to *The History of Sexuality*, as bio-power. Foucault's lectures of those years, especially 1977–8, which were not available in English until 2007, explored power in government. His lecture on governmentality was translated and available earlier, in 1991.

By 1983, but probably beginning in 1976, Foucault's (1983:208–9) work began to focus on the subject: a history of the different modes by which human beings are made subjects, especially how the individual constitutes and recognizes himself as subject. He links his corpus to the subject, and says that the subject, not power, is the theme of his life's work.[13] Foucault (1983:208) says that the subject is more important than power, but power is necessary to the study of the subject, and had to be done first; in the same way, the systematicity of knowledge probably had to be explored before it could be seen as a discursive regime.

Foucault's view of the role of the modern intellectual is activist: "The essential political problem for the intellectual is … that of ascertaining the possibility of constituting a new politics of truth … [of changing] the political, economic, institutional regime of the production of truth" (1980:133). The regime of truth can be analyzed, in Foucault's case through the analysis of historical texts, and this analysis is the first step in resisting it.

Power in Foucault[14]

I'm going to make seven points about power in Foucault that undergird the sorts of power explored in this book. The concept of power that these points outline is the first tool in what is intended as a tool-box of ideas about power in conservation.

First, Foucault's sort of power can be analyzed, but the analyses must be framed as a "how" not a "why" question: how is power exercised? (Foucault 1983: 216–17). In analyzing power in conservation, we are thus led to analyze not just discourses, but practices: through what particular practices is power exercised in conservation?

Second, power is not a structure (Foucault 1978:93). Foucault's view of power is post-structural, meaning that it is multiple and ever-changing rather than singular and stabilized into a structure.[15] Power is "the multiplicity of force relations"[16] and "the moving substrate of force relations":

> The process which, through ceaseless struggles and confrontations, transforms, strengthens, or reverses force relations … the support which these force relations find in one another, thus forming a chain or a system,[17] or on the contrary, the disjunctions and contradictions which isolate them from one another.
>
> *(Foucault 1978:92–3)*

Force relations are "unbalanced, heterogeneous, unstable, tense" (Foucault 1978:93). Capital "P" Power does not exist; it is, rather, an effect that emerges from and seeks to arrest the mobilities of power relations—and insofar as it does stop them, its success is fleeting (Foucault 1983:219–20, 1978:92–3). Foucault (1983:217, 219) increasingly prefers the term "power relations" over the singular "power."

Third, power is pervasive in society rather than located at the top of hierarchies. Power relations are "rooted deep in the social nexus," not "above"; "a society without power relations can only be an abstraction" (Foucault 1983:222–3). "The forms and the specific situations of the government of men by one another in a given society are multiple; they are superimposed, they cross, impose their own limits, sometimes cancel each other out, sometimes reinforce one another" (Foucault 1983:224). Power needs to be considered as a productive network (not a repressive hierarchy) that runs through the "whole social body" (Foucault 1980:119).

Fourth, power relations in Foucault always involve freedom. "The other," the one over whom power is exercised, must be an acting subject, a free subject; freedom must exist for power to be exerted. There is a "complicated interplay" between power and freedom: an "agonism," a "permanent provocation" (like wrestling) between power relations and freedom (Foucault 1983:221, 222n3). There are other freedoms that emerge in the 1978 lectures. First, in the second lecture, the physiocrats' edicts promote the idea that the circulation of wheat must be free, which becomes, in early French capitalism, the laissez-faire idea that the economy must be free to exercise its own nature. Second, and paralleling the economy, the population must also have its freedom to behave with minimal regulation, an idea that Foucault also traces back to the physiocrats. The freedom of the population characterizes governmentality (Foucault 1978:29–54).

Fifth, and linked to the second point above, power relations in Foucault can gel into domination, even state domination—but only for a time, because domination is always accompanied by resistance. The relation of provocation, of free play, between power relations and freedom may end with "stable mechanisms" that allow one to direct the conduct of others in a fairly constant manner. In fact, "Every strategy of confrontation dreams of becoming a relationship of power and every relationship of power leans toward the idea that … it may become the winning strategy," i.e., fixed and stable—a domination (Foucault 1983:225–6). But the same events can be interpreted as histories of struggle or of domination. Both manifest, at the level of the whole social body, the interaction between power relations and relations of freedom (Foucault 1983:225–6). Relations of power extend beyond the limits of the state (Foucault 1980:122), but they have been "progressively elaborated, rationalized, and centralized" under the auspices of state institutions (Foucault 1983:224). The state, the law, and domination are "only the terminal forms power takes" (Foucault 1978:92). In the next chapter we will see that in Foucault, discourses can be an instrument of power relations, a starting point for both dominance and resistance, "strategically codified" into a state or revolutionary ideology. Because discourses can be used by both sides, they both reinforce power and render it fragile.

Sixth,[18] power has tactics but no author. It is both intentional and nonsubjective: no power is exercised without objectives, "but this does not mean that it results from the choice or decision of an individual subject." At the local level, tactics are often quite explicit, but the comprehensive strategy they may end up forming (with other local tactics, etc.) has no creator: "The logic is perfectly clear, the aims decipherable, and yet it is often the case that no one is there to have invented them" (1978:95).

Seventh, Foucault's (1983:221) language about power shifts, from 1976 to 1983 (and in fact from 1976 to 1978), from force and struggle with a metaphor of war, to governing or conducting. Power relations involve conducting conduct, playing on the double meaning of conduct as leading (orchestra) and conduct as proper behavior (Foucault 1983:220–1). Power relations similarly involve governing, in the sense of the government of children, of families, of souls (in the case of pastoral power), or of communities. To govern is "to structure the possible field of action of others," to "act upon their actions"; the exercise of power is a "total structure of actions brought to bear upon possible actions; it incites, it induces, it seduces, it makes easier or more difficult" (Foucault 1983:220–1). This is the sort of power that informs Foucault's governmentality lecture, covered in Chapter 5.

In sum, we should ask how, not why, questions about power: how power is exercised. We should think of power relations, not capital "P" Power. Power relations are multiple and mobile, pervading society. Power always involves free subjects. It can gel into domination as a state or a revolution; it dreams of domination, but is always accompanied by resistance. Though both parties in a power relation may have intentions and explicit tactics, the larger-scale strategy that may coalesce out of these intentions and tactics in combination with others is not the result of an individual intention. Finally, in later Foucault, power becomes governing or conducting free others.

Themes in the book

A tool-box of ideas about power in conservation

I argue that Marx-based and Foucault-based approaches to understanding power in conservation are both essential—but we cannot continue to over-simplify Marx's and Foucault's ideas. This is amply demonstrated by the two ethnographies of Tania Li. One uses Foucault's idea of governmentality to understand a particular part of Sulawesi with a long history of interventions. Recently she returned to Marx, using his concept of compulsion, to understand a different part of Sulawesi without interventions. This is exactly the way it should be. The tools in this book are primarily Foucault's concepts (the power of discourses; governmentality, discipline, sovereignty, and the triangle; subject formation; and neoliberal governmentality), but also include Latour's ideas (practices of simplifying, metrology), Scott's understandings (simplification and legibility), and others. Though I don't present them as tools for thinking, most of the literature that I analyze in this book was selected to improve thinking about power in conservation. Marx enters the book through this literature, including Li.

"How" questions

Foucault (1983:216–17) asks how, not why, questions. He means "how" in the sense of "by what means is it exercised." As Agrawal (2005b:224–5) says, Foucault's ideas "prompt analyses of how problems that require government (the conduct of conduct) come into being rather than accepting unquestionably the existence of problems." Several ethnographies in this book explicitly emphasize "how" questions. This sort of question contributes significantly to avoiding over-simplifications of power. It keeps thinking focused on the details, rather than soaring to pre-determined causes. "How" questions are best answered by ethnographic methods.

Ethnography

The most productive approaches to understanding power in conservation are based on ethnographic fieldwork. I believe that understanding power and government lies in specifics, not in generalizations. There are several domains with high relevance to effective conservation that are only accessible through ethnographic research, including project implementation and household-level economies. Some non-ethnographic methods, particularly history, can produce a similar wealth of qualitative data at a similar level. Foucault's own theory is based on the detailed study of historical texts. It is in many ways comparable to ethnography, except that it cannot produce understandings of ideas *in practice*—and ethnographies can.

Economy and power

We need to use Foucault to question, not assume, the role of the economic in power. Foucault never sees power and the economy as separate things. Some of his most interesting writing analyzes economists' thinking about government. We should not reduce the political to the economic. "Large-scale economic forces" need to be broken down analytically to a level of detail that will reveal the actual interaction between, for example, political clout and financial gain, or vice versa. There are winners and losers, but we can analyze the process that yields them, and it is not the same for everyone, everywhere—and today's winners may be tomorrow's losers. Capitalism itself needs to be broken down and studied in particular places at particular times.

Ecology

My favorite ethnographies fully realize the influence of ecology. So, for example, the ecologies of cacao and rainforest soils contribute to the way that Marx's concept of compulsion plays out in highland Sulawesi, Indonesia, as analyzed by Tania Li. Timothy Mitchell describes the role of curly pond weed, *Anopheles* mosquitos, and the malaria parasite in Egypt's disasters of the 1940s, all enabled by a dam and irrigation. Stepping back from the influence of ecology to ecology as a science brings power in. Paul Nadasdy (2003), for example, nicely contrasts First Nations peoples' understandings of ecology with ecological science, which separates species and counts animals. And Anna Tsing (2003) contrasts the cultural and social aspects of honey hunting in Kalimantan, Indonesia, with an ecology that sees only forests.

The organization of the book

I am an environmental anthropologist. But my discipline is not exclusionary, so this book also examines works written by historians (especially William Cronon and Richard Grove), geographers (especially Roderick Neumann and Bruce Braun), and political scientists (especially Arun Agrawal, James C. Scott, and Timothy Mitchell). I selected material almost exclusively for its importance to the understanding of power in conservation. Most pieces are also about a particular place (the exceptions are Scott, who did however carry out field-based research in Malaysia, and Haraway). Ethnographies include the United States (Cronon, Haraway, some Tsing, some Mitchell), Canada (Nadasdy, Braun), India (Agrawal, Grove, Mosse), Latin America (Escobar, Gudeman, Lazar, Mitchell), Africa (Bonneuil, Neumann, Moore, Fairhead and Leach, Ferguson), Indonesia (Ellen, Li, Tsing), and others (Mitchell—Egypt).

Chapters 2, 5, 7, 9, and 11 are primarily theoretical chapters. Chapters 2, 5, 9, and 11 examine four concepts in Foucault's theory that lie behind the environmental anthropology literature on power analyzed in the remaining chapters, and are thus important to an accurate understanding of this literature. The four ideas are described above. Alongside Foucault's general ideas of power as outlined, I intend these concepts to be used as tools for improving the effectiveness of conservation. Chapter 7 examines ideas from Bruno Latour and James Scott that also often appear in the literature on power in conservation, offset by an ethnographic piece by Anna Tsing. In *Seeing Like a State* (1998), Scott argues that states simplify or render legible in order to govern, and that modern states base their simplifying interventions on a valorization of science and technology. Latour is talking about science, not states. He adds the actual technologies and practices of simplifying, rendering legible, and calculating, and how they extend power across space, to Scott. Latour's and Scott's ideas can also be added to the tool-box offered in this book. Tsing's ethnography provides an example of the complexity and potential of ethnography to set against state and scientific simplification of relations between people and forests.

Chapters 3, 4, 6, 8, 10, and 12 all use ethnography to elucidate and apply the theoretical chapters they follow. Chapter 3 introduces four authors that brought Foucault to environmental anthropology, particularly Foucault's concept of the power of discourses: James Ferguson, Arturo Escobar, and James Fairhead and

Melissa Leach. Chapter 4 analyzes two pieces by the historian William Cronon that reveal the power of discourses in Western (or Western-influenced) conservation. It also examines an article by Roy Ellen in which an indigenous people change their understanding of the forest and their identity from supporting the logging industry to a conservationist position. Chapter 6 provides four ethnographies of the triangle of sovereignty, discipline, and governmentality, by Roderick Neumann, Christophe Bonneuil, Donald Moore, and Tania Li. The chapter applies these ideas to conservation, and also discusses relations within the triangle in conservation. Chapter 8 returns to knowledges to consider articulations between Western (global) and indigenous (local) knowledges, articulations that reveal the play of power between environmental knowledges. The chapter examines Richard Grove's history of a colonial–local articulation that shaped the science of botany, using Foucault and Latour. Tania Li's ethnography considers the implications of the creation of a body of indigenous environmental knowledge to claim land rights. Paul Nadasdy uses Latour to analyze a project to integrate scientific and traditional environmental knowledge for environmental co-management in the Yukon. In Chapter 10 Bruce Braun, Arun Agrawal, and Tania Murray Li all use Foucault's ideas about how subjects are created to explore their ethnographies. Braun analyzes the historical impact of the science of geology. Agrawal and Li examine conservation projects that were actually designed to create conservation subjects, with some success in Agrawal's case study, but a fairly dramatic failure in Li's. In Chapter 12, I examine a failed project to add subject formation to micro-lending groups in Bolivia studied by Sian Lazar; a large-scale project in Indonesia to make communities into development subjects analyzed by Tania Li; the odd phenomenon of non-market neoliberal subjects explored by Tania Li; and subject creation aspects of the contemporary global economy that utilize cultural differences examined by Anna Tsing. Subject formation pervades conservation and development, with a long history of user groups, social capital, microcredit, and community-driven development. It also pervades our economy, especially its low-wage sectors. In both locations, it is neoliberal, based on a core assumption that poor people can be improved by guiding them to become *homo economicus*. But these readings suggest that neoliberal subject creation depends on the non-economic: culture and society.

Chapters 13 to 17 explore topics in environmental anthropology that have been opened up by Foucault or Latour. Marx also re-enters the book here, in uses that recognize the complexity of his ideas. In Chapter 13, I use ethnographies, by the economic anthropologist Stephen Gudeman, the political scientist Timothy Mitchell (influenced by Foucault and Latour), and Tania Li (using Marx), to question the economic perspective that pervades conservation interventions. In Chapter 14, I compare David Mosse's idea that the implementation of conservation and development projects is invisible to policy-makers; James Scott's idea that mētis, informal and local knowledge and practices, is devalued by states but essential to the success of state projects; and Tania Li's thesis that the compromises that characterize project implementation are integral to state power. Chapter 15 considers assemblages. I use Tania Li, whose analysis uses Foucault, and Timothy Mitchell, whose analysis uses Latour, to examine practices of assemblage. I argue that when these assemblages are successful and an intervention occurs, the *effects* of the intervention, typically unplanned and unanticipated, also constitute an assemblage, and use Mitchell to illustrate assemblages of effects that include the non-human. In Chapter 16, I explore Anna Tsing's ideas about universals and collaborations. I compare this to Annelise Riles' analysis of the patterned process of creating a United Nations (UN) agreement. Both universals and global agreements are meaningless. And the collaborations they inspire are characterized by misunderstandings rather than the consensus we expect. This chapter is highly relevant to conservation discourses, collaborations, and global agreements. The main universal Tsing explores is capital "N" Nature, and her examples are thus all environmental. Riles' UN agreement involves gender, but international environmental agreements are negotiated in much the same way. In Chapter 17, I explore an emerging literature that uses natural history to tell stories of assemblages that include humans and non-humans making worlds in collaboration: world-making. Such stories may help us rethink survival in the Anthropocene, a term in common and informal scientific usage for the current geological age, when humans have dominated the climate and the environment. It is a term that unfortunately enshrines the degradation story, and the opposition between people (anthropo-) and the earth assumed by that story. What if the story were changed?

Notes

1. IUCN was founded in 1948, followed by the World Wide Fund for Nature (WWF) in 1961, the UN Environmental Programme in 1972, World Resources Institute in 1982, and Conservation International in 1987. The New York Zoological Society, which became the Wildlife Conservation Society, was founded in 1895, and carried out international research very early, but its first intervention was in 1969.
2. I should point out that community forestry models are typically brought in when governments' forest departments have failed to preserve forests in state reserves.
3. Actually material production and economic relations.
4. The superstructure includes legal and cultural dimensions of society.
5. Williams' essay was originally published in the *New Left Review* in 1973.
6. I would add Adams and Hutton 2007, which specifically addresses political ecology and conservation, focusing on parks. They combine political economy (66ff.) with a consideration of the Foucault-influenced "idea of pristine nature" (152ff.).
7. I use Robbins here because his book is intended as a textbook, is fairly recent, and actually incorporates several of Foucault's ideas.
8. In Foucault's thought there is no "larger social engine," power is based on freedom, not control, and profit and extraction are less significant than reliance on the economy to govern.
9. Foucauldian approaches to political ecology emerged in the late 1990s (Watts 2019:34). Bryant 1998 (82) argues that this shift was part of a "second phase" in political ecology, developing out of "concerns over the influence of deterministic neo-Marxism," when political ecologists began drawing upon an "eclectic" range of theories. The first was

Peet and Watts' 1996 (37) *Liberation Ecologies* (which coined the term "liberation ecology" for the combination of Marxist political ecology with the post-structural idea that truth is socially constructed), which includes Escobar's chapter explicitly calling for an integration of political ecology and Foucauldian ideas (see also Escobar 1999). Fletcher 2010 also argues for a more Foucauldian political ecology (which he calls, after Peet and Watts, liberation environmentality) to counter neoliberal governmentality. Most recently, *The Routledge Handbook of Political Ecology* (Perreault, Bridge, and McCarthy 2019), like Robbins, simply incorporates several of Foucault's ideas about power. Discursive power ("truth," power/knowledge) was the first of Foucault's concepts of power to be used by political ecologists, followed by subject formation and governmentality.

10. Agrawal (2005b:224–5) adds: Foucault's ideas "prompt analyses of how problems that require government (the conduct of conduct) come into being rather than accepting unquestionably the existence of problems." They do not just identify causes of degradation or propose solutions. They ask "when and for what reasons these processes came to be identified as problems that merit a particular style of analysis and resolution." He also adds attention to the techniques, forms, and representations of knowledge related to government, and questions about the relation between government and self-construction.
11. Foucault (1983:208–9, 1980:109–33) himself says this, linking all three to the subject (Bernauer and Keenan 1987:165).
12. Examples from this period include the 1970[1966] *The Order of Things* and the 1972[1969] *The Archeology of Knowledge*.
13. In this period, *The Care of the Self*, Volume 3 of *The History of Sexuality*, was published (1986[1984]); in 1983 he was lecturing on the government of the self.
14. This section of the Introduction draws on Foucault 1983. It was written after his governmentality lecture, and shares language with it. It helps clarify relations between power relations, discursive power (Chapter 2) and governmentality (Chapter 5).
15. In 1968 Foucault (1991:72) wrote in a footnote: "Is it necessary to point out yet again that I am not what is called a 'structuralist'?"
16. Note the use of the work "force" here rather than "power," a usage that may have been picked up by political ecologists like Robbins—but Foucault abandoned the use of this term. See my seventh point under "Power in Foucault" later in this Introduction.
17. This clause sounds like Bruno Latour's networks or Timothy Mitchell's assemblages. See Chapter 15.
18. This draws from Foucault 1978, covered in more depth in the next chapter.

References

Adams, William M. and Jon Hutton 2007 People, Parks and Poverty: Political Ecology and Biodiversity Conservation. *Conservation and Society* 5(2):147–83.

Adger, W. Neil, Tor A. Benjaminsen, Katrina Brown, and Hanne Svarstad 2001 Advancing a Political Ecology of Global Environmental Discourses. *Development and Change* 32: 681–715.

Agrawal, Arun 2005a Environmentality: Community, Intimate Government, and the Making of Environmental Subjects in Kumaon, India. *Current Anthropology* 46(2):161–90.

Agrawal, Arun 2005b *Environmentality: Technologies of Government and the Making of Subjects.* Durham, NC: Duke University Press.

Baylis, Kathy, Jordi Honey-Rosés, Jan Börner, Esteve Corbera, Driss Ezzine-de-Blas, Paul J. Ferraro, Renaud Lapeyre, U. Martin Persson, Alex Pfaff, and Sven Wunder 2016 Mainstreaming Impact Evaluation in Nature Conservation. *Conservation Letters* 9(1):58–64.

Baynes, Jack, John Herbohn, Carl Smith, Robert Fisher, and David Bray 2015 Key Factors which Influence the Success of Community Forestry in Developing Countries. *Global Environmental Change* 35:226–38.

Bernauer, James and Thomas Keenan 1987 Michel Foucault: A Biographical Chronology. *Philosophy and Social Criticism* 12:270–7.

Biersack, Aletta and James B. Greenberg, eds. 2006 *Reimagining Political Ecology*. Durham: Duke University Press.

BirdLife International 2018 *State of the World's Birds: Taking the Pulse of the Planet*. Cambridge, UK: BirdLife International.

Blaikie, Piers and Harold Brookfield. 1987 *Land Degradation and Society*. London: Methuen.

Börner, Jon, Kathy Baylis, Esteve Corbera, Driss Ezzine-de-Blas, Paul J. Ferraro, Jordi Honey-Rosés, Renaud Lapeyre, U Martin Persson, and Sven Wunder 2016 Emerging Evidence of the Effectiveness of Tropical Forest Conservation. *PLoS One* 11(11):1–11.

Börner, Jan, Kathy Baylis, Esteve Corbera, Driss Ezzine-de-Blas, Jordi Honey-Rosés, U Martin Persson, and Sven Wunder 2017 The Effectiveness of Payments for Environmental Services. *World Development* 96:359–74.

Bowler, Diana E., Lisette M. Buyung-Ali, John R. Healey, Julia P.G. Jones, Teri M. Knight, and Andrew S. Pullin 2012 Does Community Forest Management Provide Global Environmental Benefits and Improve Local Welfare? *Frontiers in Ecology and the Environment* 10(1):29–36.

Bryant, Raymond L. 1998 Power, Knowledge and Political Ecology in the Third World: A Review. *Progress in Physical Geography* 22(1):79–94.

Bryant, Raymond L., ed. 2015 *International Handbook of Political Ecology*. Cheltenham, UK: Edward Elgar.

Calhoun, Craig, ed. in chief 2002 *Dictionary of the Social Sciences.* New York, Oxford University Press.

Escobar, Arturo 1996 Constructing Nature: Elements for a Poststructural Political Ecology. In *Liberation Ecologies: Environment, Development, Social Movements*. R. Peet and M. Watts, eds. Pp. 46–68. London: Routledge.

Escobar, Arturo 1999 After Nature: Steps to an Antiessentialist Political Ecology. *Current Anthropology* 40(1):1–30.

Fisher, Brendan, Andrew Balmford, Paul J. Ferraro, Louise Glew, Michael Mascia, Robin Naidoo, and Taylor H. Ricketts 2013 Moving Rio Forward and Avoiding Ten More Years with Little Evidence for Effective Conservation Policy. *Conservation Biology* 28(3):880–2.

Fletcher, Robert 2010 Neoliberal Environmentality: Towards a Poststructuralist Political Ecology of the Conservation Debate. *Conservation & Society* 8(3):171–81.

Foucault, Michel 1970[1966] *The Order of Things: An Archaeology of the Human Sciences.* New York: Vintage Books.

Foucault, Michel 1972[1969] *The Archaeology of Knowledge*. New York: Pantheon.

Foucault, Michel 1977[1975] *Discipline and Punish: The Birth of the Prison*. Alan Sheridan, trans. New York: Vintage Books.

Foucault, Michel 1978[1976] *The History of Sexuality. An Introduction: Volume 1.* Robert Hurley, trans. New York: Vintage Books.

Foucault, Michel 1980 *Power/Knowledge: Selected Interviews and Other Writings 1972–1977.* Colin Gordon, ed. New York: Pantheon Books.

Foucault, Michel 1983 Afterword: The Subject and Power. In *Michel Foucault: Beyond Structuralism and Hermeneutics*. 2nd edition. Hubert L. Dreyfus and Paul Rabinow, eds. Pp. 208–26. Chicago: The University of Chicago Press.

Foucault, Michel 1991[1968] Politics and the Study of Discourse. In *The Foucault Effect: Studies in Governmentality with Two Lectures by and an Interview with Michel Foucault*. Graham Burchell, Colin Gordon, and Peter Miller, eds. Pp. 53–72. Chicago: The University of Chicago Press.

Foucault, Michel 1994[1974]. Prisons et asiles dans le mécanisme du pouvoir. C. O'Farrell, trans. *Dits et Ecrits* 11:523–4.

Geldmann, Jonas, Megan Barnes, Lauren Coad, Ian D. Craigie, Marc Hockings, and Neil D. Burgess 2013 Effectiveness of Terrestrial Protected Areas in Reducing Habitat Loss and Population Declines. *Biological Conservation* 161:230–8.

Greenberg, James B. and Thomas K. Park. 1994 Political Ecology. *Journal of Political Ecology* 1:1–12.

Guerrero, Angela M., Nathan J. Bennett, Kerrie A. Wilson, Neil Carter, David Gill, Morena Mills, Christopher D. Ives, Matthew J. Selinske, Cecilia Larrosa, Sarah Bekessy, Fraser A. Januchowski-Hartley, Henry Travers, Carina A. Wyborn, and Ana Nuno 2018 Achieving the Promise of Integration in Social-ecological Research: A Review and Prospectus. *Ecology and Society* 23(3):38–65.

IUCN n.d. *IUCN Red List*. Accessed 7/23/2019. www.iucnredlist.org/

Mascia, Michael B., J. Peter Brosius, Tracy A. Dobson, Bruce C. Forbes, Leah Horowitz, Margaret A. McKean, and Nancy J. Turner 2003 Editorial: Conservation and the Social Sciences. *Conservation Biology* 17(3):649–50.

Moore, Donald S. 1996 Marxism, Culture, and Political Ecology: Environmental Struggles in Zimbabwe's Eastern Highlands. In *Liberation Ecologies: Environment, Development, Social Movements*. R. Peet and M. Watts, eds. Pp. 125–47. London: Routledge.

Nadasdy, Paul 2003 *Hunters and Bureaucrats: Power, Knowledge, and Aboriginal–State Relations in the Southwest Yukon*. Vancouver, BC: University of British Columbia Press.

Peet, Richard and Michael Watts, eds. 1996 *Liberation Ecologies: Environment, Development, Social Movements*. London: Routledge.

Peet, Richard and Michael Watts, eds. 2004 *Liberation Ecologies: Environment, Development, Social Movements*. 2nd Edition. London: Routledge.

Perreault, Tom, Gavin Bridge, and James McCarthy, eds. 2019 *The Routledge Handbook of Political Ecology*. London, Routledge.

Robbins, Paul 2012 *Political Ecology: A Critical Introduction*. 2nd Edition. London: Wiley-Blackwell.

Scott, James C. 1998 *Seeing Like a State: How Certain Schemes to Improve the Human Condition Have Failed*. New Haven: Yale University Press.

Silvey, Rachel 2010 Rethinking Power and Development: A Review Forum on Tania Murray Li's *The Will to Improve*. *Annals of the Association of American Geographers* 100(1):222–3

Tsing, Anna L. 2003 Cultivating the Wild: Honey-hunting and Forest Management in Southeast Kalimantan. In *Culture and the Question of Rights*. Charles Zerner, ed. Pp. 24–55. Durham, NC: Duke University Press.

Vayda, Andrew P. and Bradley B. Walters 1999. Against Political Ecology. *Human Ecology* 27(1):167–79.

Vayda, Andrew P. 2009 Causal Explanation as a Research Goal: Dos and Don'ts. In *Explaining Human Actions and Environmental Changes*. A.P. Vayda, ed. Pp. 1–47. Lanham, MD: AltaMira Press.

Vidal, John 18 September 2019 We're Losing Species at Shocking Rates—So Why Is Conservation Failing? *Guardian*.

Watts, Michael J. 2019 Now and Then: The Origins of Political Ecology and the Rebirth of Adaptation as a Form of Thought. In: *The Routledge Handbook of Political Ecology*. T. Perreault, G. Bridge, and J. McCarthy, eds. Pp. 19–50. London, Routledge.

Williams, Raymond 2005[1980] Base and Superstructure in Marxist Cultural Theory. In *Culture and Materialism: Selected Essays.* R. Williams, ed. Pp. 31–49. London: Verso.
World Bank. 2019 *The World Bank Data*. https://data.worldbank.org/indicator/AG.LND.FRST.ZS.
Zimmerer, Karl S. and Thomas J. Bassett, eds. 2003. *Political Ecology: An Integrative Approach to Geography and Environment-Development Studies*. New York. The Guilford Press.

2

DISCOURSES AND THEIR POWER IN FOUCAULT

Introduction

The power of discourses is the second of Foucault's ideas about power that needs to be in the tool-box of those committed to effective conservation. Conservation is full of discourses in Foucault's sense, as we will explore further in the next two chapters: nature is one, and wilderness, conservation itself, the environment, sustainability, biodiversity, endangered species, etc. How did this rich discursive field come to gather around and join in the huge, complex project of conservation interventions in the developing world? And how has this discursive field and its power influenced the effectiveness of conservation in the real world? These sorts of questions cannot begin to be answered without a robust understanding of what discourses and their power are.

We have seen that the idea of the power of knowledge, compared to Foucault's other concepts, had the earliest and largest impact on political ecology. However, the relationship between discourses or knowledge and power in political ecology can be over-simplified such that discourses are seen as created by the powerful, serving the interests of those with political-economic power, or working to increase the power of the powerful. This is simply not the way Foucault understands either discourses or power. Discourses emerge rather like trends; they are not created; they are authorless. They are also not wielded nor controlled by those with power (power itself is not wielded nor controlled); discourses can serve both states and resistances. This means that they introduce a fragility to the power of states or other players.

As we have seen in Chapter 1, discourses or knowledge was not always linked to power in Foucault's thought: in his first books he focused on the systematicity of knowledge rather than discursive *regimes*. This chapter follows the relation between discourses and power through Foucault's writing, beginning with the late 1960s and ending in the late 1970s. This allows us to begin with the period *before* power and trace the development of his thinking from the first revelation of power to an endpoint in which discourses and power have become inextricable.

This serves as an appropriate introduction to Foucault's corpus, which, as I have said in Chapter 1, is not static in its language or its ideas. As Foucault thinks through the concept of discourse, the term becomes pluralized as "discourses" or "discursive practices," then generalized as "knowledge" and fragmented as "discursive elements." As discourses are things that have come to be accepted as true, "truth" and "scientific truth" also often appear as synonyms for powerful discourses. They are all synonyms in Foucault's writing in a sense, but certain changes—for example, the shift from discourse to discourses—signal important developments in his thought.

Discourse before power: Foucault's thought in the late 1960s

In the late 1960s, discourses do not yet have power in Foucault's thinking, but he sees his thought about discourses as political.

In a 1968 response to an interview question,[1] Foucault (1991[1968]:62) writes that the goal of his work is to "liberate the discursive field from the historical-transcendental structure which nineteenth century philosophy imposed on it." In this piece, entitled "Politics and the Study of Discourse," Foucault's (1991:65) politics consist of the questioning, and thus the "meticulous destruction" of this structure. In this nineteenth century structure, "the history of the mind, consciousness, reason, knowledge, ideas or opinions" explains human progress, while discursive practice is reduced to "a shallow transparency," "just air, extraneous matter, a flutter of wings scarcely audible among the earnestness of history and the silence of thought" (Foucault 1991:54, 65). He characterizes this approach to history as "totalizing history" (Foucault 1991:54). This understanding of the progress of human history supports "the status … the conditions of exercise, functioning and institutionalization of scientific discourses" (Foucault 1991:65). Foucault (1991:65) sees this—scientific truth—as an important problem for political practice. To question scientific truth with political intent, he re-analyzes historical discursive practices to argue against nineteenth century philosophy.

Foucault's own summary of what he has been doing for the past ten years (from working on his doctorate to just before his election to the College de France), which he puts in single quotation marks, is:

> to determine … what the mode of existence of discourses and particularly scientific discourses … must have been in Europe, since the

> seventeenth century, in order that the knowledge which is ours today could come to exist … particularly that knowledge which has taken as its domain this curious object which is man.
>
> *(Foucault 1991:70)*

He has been making what he calls "differentiated analyses" of European scientific discourses beginning in the seventeenth century, particularly the discursive practices of the human sciences, in an effort to liberate them from the "totalizing history" of nineteenth century philosophy, which emphasized "mind, consciousness, reason, knowledge, ideas or opinions" over discourse (Foucault 1991:54).

To define discourses, Foucault (1991:54) begins with a problem: "The problem I have set myself is that of the individualization of discourses," those "curious entities" that one speaks of in the singular, and which "one believes one can recognize at first glance, but whose limits one would have some difficulty in defining." In other words, the problem is: what are these discourses that become, curiously, individualized or singular? What is one speaking of when one uses these words? He gives many examples: *system* (a term used by his interviewer in this piece), ancient discourses like *medicine* and *mathematics*, recent ones like *economics* or *psychiatry*, ones like *casuistry* that have disappeared, and discourses like *sociology* or *psychology* that have been undergoing constant change. He also mentions *political economy*, *general grammar*, *natural history*, *psychopathology*, *clinical medicine*, *biology*, and *chemistry* (Foucault 1991:54). Many of these examples are academic disciplines, in keeping with Foucault's political intent of questioning scientific truth.

Foucault offers three criteria to define a discourse. First, a discourse is "singular" when one can formulate a set of rules of formation for all the discourse's objects, operations, concepts, and theoretical options—in spite of the lack of coherence between them. Second, one can define an historical moment when new rules of formation came into effect, a moment of transformation, of discontinuity. And third, one can define a set of relations that situate the discourse among other discourses and in the non-discursive context (e.g., institutions) (Foucault 1991:54). Restated several pages later: "In discourse something is formed, according to clearly definable rules … this something exists … changes, disappears," in a relationship with everything a society can produce; in other words, "There is the formation and transformation of 'things said'. It is the history of these 'things said' that I have undertaken to write" (Foucault 1991:63). Restating the question the piece is answering (see Note 1 for the question), Foucault (1991:61) says that his work is "an attempt to introduce 'the diversity of systems and the play of discontinuities into the history of discourses.'"

Foucault's (1991:59–60) method of the differentiated analysis of discursive practices is an "archeology," "the description of an archive," by which he means unearthing "the set of rules which at a given period and for a given society" define: what can be said, which utterances are circulated and which repressed, which utterances are considered valid and which invalid, which previous discourses are retained and what role are they given, and how the struggle to control discourses is conducted. Foucault (1991:55) has "studied … ensembles of discourse: I have characterized them; I have defined the play of rules, of transformations, of thresholds, of remanences [what remains or is residual from an earlier period]. I have collated different discourses and described their clusters and relations." There are three consequences of this archeological method, which repeat the three parts of his definition: so the defining singularity of a discourse means we should treat it as a "*monument* to be described in its intrinsic configuration," the defining moment when a discourse emerges means we should investigate its "conditions of existence," and finally the defining relations between discourses and their non-discursive context means we should relate a discourse to "the practical field in which it is deployed" (Foucault 1991:60–1, author's italics).

In this early period, Foucault makes a final point about discourses that will continue to inform his later thinking about power: that discoursing subjects are *not* the sources of discourses.[2] Rather, "Discoursing subjects form a part of the discursive field—they have their place within it" (Foucault 1991:58). Foucault's method is "to cut up, analyze, combine, recompose all these texts so that now the transfigured face of their author is never discernible." This is difficult to accept for two reasons: because people do not like to think that "their history, their economy, their social practices, the language they speak, their ancestral mythology, even the fables told them in childhood, obey rules which are not given to their consciousness," and because everyone "hopes and believes he put something of 'himself' into his own discourse" (Foucault 1991:71). For Foucault, people are not the sources of discourses; history, economy, society, language, culture, and the discourses that come out of their own mouths are not their own. He ends the interview on "Politics and the Study of Discourse" by saying:

> In each sentence that you pronounce—and very precisely in the one that you are busy writing at this moment, you who have been so intent, for so many pages, on answering a question in which you felt yourself personally concerned and who are going to sign this text with your name—in every sentence their [*sic*] reigns the nameless law, the blank indifference: "What matter who is speaking; someone has said: what matter who is speaking."
>
> *(Foucault 1991:72)*

Here Foucault quotes Samuel Beckett. This particular theme remains unchanged as Foucault's concept of power develops. In other words, the subject as author will be just another part of the field of the power of discourses, as he is in 1968 just another part of the discursive field.

If Foucault has not seen power in discourses in 1968, he has seen the political potential of analyzing discourses. In the language of this piece, nineteenth century philosophy "imposed" a historical-transcendental structure on the discursive field, and his own philosophy intends to liberate it, by, as he writes, "trying to define how, to what extent, and at what level discourses, particularly scientific discourses, can be objects of a political practice, and in what system of dependence they can exist in relation to it" (Foucault 1991:69). This is as close as he gets to the power of discourses in this piece. But much of his thinking about discourses, including the fact that they are authorless, does not change after the revelation that discourses have power.

The revelation of power

In Foucault's writings of the late 1970s, which are often cited by political ecologists and others, discourse can be instrument and effect of power, it can transmit, produce, and reinforce power—or undermine it. Foucault is saying that power relations and discourses are intimately connected and mutually dependent, but he is thinking about the relation between discourse and power by thinking of them separately, as though there are discourses and there are power relations, two different things that interact. This makes it easy to misread Foucault in this period. Ultimately he makes it clear that power does not act on but is rather internal to knowledge. In fact, knowledge governs, like governmentality (which he has already begun to think about but not written about): it structures the field of possibilities.

Foucault's revelation of power informed *The History of Sexuality: An Introduction*, published in French in 1976 and in English in 1978, and I believe he worked out the idea that knowledge is imbued with power in that book. *Power/Knowledge* is an edited volume that covers Foucault's writings between 1972 and 1977, bridging the period of Foucault's first revelation of power and his final understanding of the relation between power and discourse. In 1977 Foucault is just beginning to think about the subject; he will give his lecture on governmentality in one year.

As I have said in Chapter 1, Foucault's language about power shifts from force and struggle with a metaphor of war to governing or conducting (1983:221, 1980:123). In *The History of Sexuality*, the word "power" occurs, but it is defined through the term "force relations." In *Power/Knowledge*, discourse and knowledge are synonyms (e.g., "discourse and forms of knowledge"), with truth as another synonym (Foucault 1980:112).

Foucault's view of discourses, like his view of power, becomes increasingly post-structural. By 1976, "discursive practices" become "discursive elements." We do not have, he writes, a dominant discourse versus a dominated one, but a multiplicity of discursive elements:

> We must not imagine a world of discourse divided between accepted discourse and excluded discourse, or between the dominant discourse and the dominated one; but as a multiplicity of discursive elements that can come into play in various strategies. It is this distribution that we

> must reconstruct, with the things said and those concealed, the enunciations required and those forbidden, that it comprises.
>
> *(1978:100)*

And again,

> There is not, on the one side, a discourse of power, and opposite it, another discourse that runs counter to it. Discourses are tactical elements or blocks operating in a field of force relations; there can exist different and even contradictory discourses within the same strategy; they can, on the contrary, circulate without changing their form from one strategy to another, opposing strategy.
>
> *(1978:101–2)*

Discursive elements do not simply serve power. They serve force (later power) relations, including both the side dreaming of dominance and the side resisting that dominance. Those we tend to see as having political and/or economic power have strategies that make use of discursive elements, but so do those who resist them—and sometimes they are the same elements. Discursive elements circulate between them in all their mobility, lending themselves to resistance as well as to dominance. Thus, discourse produces and reinforces power, but also "renders it fragile" as it produces and reinforces resistance (Foucault 1978:101). Resistances are as multiple, shifting, and local as force relations; they can be "strategically codified" into a revolution, as force relations can be into a state (Foucault 1978:96).

The existence of tactics and strategies does not reverse Foucault's 1968 point that discoursing subjects are not the source of discourses. When Foucault first "sees" power, he sees it as being like discourses in not having its source in an author. Power, he writes, is both intentional (strategic, tactical) and non-subjective; no power is exercised without objectives, "but this does not mean that it results from the choice or decision of an individual subject" (Foucault 1978:95). Multiple and often explicit local tactics form a larger-scale strategy together; this grand strategy has no inventor (see Chapter 1).

In *Power/Knowledge*, Foucault gives an example of a coherent strategy, with a variety of tactics including discourses, which has no author: fixing workers at their places of work in France in 1825–30. Tactics include: pressuring workers to marry, providing free housing, credit slavery (he references Marx), savings-bank systems, grocers and wine merchants who act for the bosses, parliamentary measures for the schooling of children, and local chambers of commerce (Foucault 1980:202–3). Discourses are also tactics in this strategy: "Around all this there is formed little by little a discourse, the discourse of philanthropy and the moralisation of the working class" (Foucault 1980:204). He writes, of the strategy: "You get a coherent, rational strategy, but one for which it is no longer possible to identify a person who conceived it" (Foucault 1980:203). "So the objective existed and the strategy was developed, with ever-growing coherence, but without it being necessary to attribute to it a subject which makes the law, pronouncing it in the form of 'Thou shalt' and 'Thou shalt not'" (Foucault 1980:204). The interviewer asks about class, and Foucault (1980:203) says that there is a reciprocal relation between class and the strategy, that one cannot say that the bourgeois class "invented and forcibly imposed this strategy on the working class." In this example, the discourse of philanthropy and the moralization of the working class "formed little by little" around a strategy to fix workers at their place of work as that authorless strategy cohered.

Similarly, in *The History of Sexuality*, Foucault (1978:97) examines "the discourses of truth" that "have taken charge" of sex, and their relation to power relations. The question to study, he writes, is:

> In a specific type of discourse on sex … appearing historically and in specific places (around the child's body … and so on), what were the most immediate, the most local power relations at work? How did they make possible these kinds of discourses, and conversely, how were these discourses used to support power relations? … How were these power relations linked to one another according to the logic of a great strategy, which in retrospect takes on the aspect of a unitary and voluntarist politics of sex? In general terms: rather than referring all the infinitesimal violences that are exerted on sex, all the anxious gazes that are directed at it … to the unique form of a great Power, we must immerse the expanding production of discourses on sex in the field of multiple and mobile power relations.
>
> *(Foucault 1978:97–8)*

In this example, discourses on sex, which appear historically and surround a particular "place" like children's bodies, are produced and expand "in the field" of "multiple and mobile power relations," all local vis-à-vis children's bodies. The discourses are made possible by and support power relations. The two act on each other.

Similarly, in a 1976 lecture appearing in *Power/Knowledge*, Foucault (1980:93) writes:

> In any society, there are manifold relations of power which permeate, characterize and constitute the social body, and these relations of power cannot themselves be established, consolidated nor implemented without the production, accumulation, circulation and functioning of a discourse.… We are subjected to the production of truth through power and we cannot exercise power except through the production of truth.

Here, too, power relations and discourses act on each other.

Focusing on the context of power relations out of which a specific discourse on sex emerges produces four suggested methods, of which the fourth and the first are most relevant here.[3] The fourth rule is the tactical polyvalence of discourses. As we saw above:

> Discourse can be both an instrument and an effect of power, but also a hindrance … a point of resistance and a starting point for an opposing strategy. Discourse transmits and produces power; it reinforces it, but also undermines and exposes it, renders it fragile and makes it possible to thwart it.
>
> *(Foucault 1980:101)*

Here discourse can be instrument and effect of power, it can transmit, produce, and reinforce power—and also the opposite. He is emphasizing the way that discourse acts on power in this rule, but he is doing so to show that discourse can undermine as well as support power.

The first rule is something new, the rule of immanence: "Between *techniques* of knowledge and strategies of power, there is no exteriority" (1978:98, author's italics). This rule anticipates power/knowledge. In fact, he uses the term "power-knowledge" here, in 1976, in *The History of Sexuality* (Foucault 1978:98). This immanence is very clear in a 1977 interview in *Power/Knowledge* called "Truth and Power."

In this 1977 interview, in answer to a question about discontinuity in history, Foucault (1980:112–13) thinks back to a time before his revelation of power:

> It's not so much a matter of knowing what external power imposes itself on science, as of what effects of power circulate among scientific statements, what constitutes … their internal régime of power….
>
> It was these different régimes that I tried to identify and describe in *The Order of Things*…. But what was lacking here was this problem of the "discursive régime" … I confused this too much with the systematicity, theoretical form, or something like a paradigm.

He rewords his original 1968 definition of discourse (clearly definable rules): "It is a question of what *governs* statements, and the way in which they *govern* each other so as to constitute a set of propositions which are scientifically acceptable…. In short, there is a problem of the régime, the politics of the scientific statement" (1980:112, author's italics). Knowledge governs, in the way he has already begun to think about: it structures the field of possibilities, like governmentality.

This same idea is also stated in terms of truth. What matters to Foucault (1980:118) is not

> drawing the line between that in a discourse which falls under the category of scientificity or truth, and that which comes under some other category [i.e., ideology], but in seeing historically how effects of truth are produced within discourses which in themselves are neither true nor false.

Here "effects of truth" constitute the power of discourses: "Truth isn't outside power, or lacking in power," "it induces regular effects of power." "Each society has its régime of truth, its 'general politics' of truth: that is, the types of discourse which it accepts and makes function as true" (Foucault 1980:131). Truth is "the ensemble of rules according to which the true and the false are separated and specific effects of power attached to the true" (Foucault 1980:132). "'Truth' is linked in a circular relation with systems of power which produce and sustain it, and to effects of power which it induces and which extend it. A régime of truth." (Foucault 1980:133). In fact, "Truth is already power" (Foucault 1980:133).

To sum up, by the mid-1970s, Foucault has realized that a discourse seen as "truth" in any specific time and place—knowledge—has power; power is immanent in and internal to knowledge. He has recognized that "the historical-transcendental structure imposed on the discursive field" of 1968 was knowledge with truth attached.

Conclusion

Foucault's understanding of knowledge, including science, as political, was hugely important in the social science of conservation, and, as we have seen, in political ecology, beginning in the mid-1990s. However, it is often over-simplified.

Michel Foucault sees power in the knowledge that is accepted as truth, especially scientific truth. He says that every society has its truth regimes, which separate what will be considered true and what will be considered false and attaches power to the true. In other words, considering something to be true privileges it with power. We are typically not conscious of these truth regimes; for the most part, people simply believe that what is defined as truth *is* truth. For Foucault, what is defined as truth is not "true"—it is knowledge that has come to be considered truth in a political process (the tactics and strategies of power relations), and the effect of this process is that what is defined as truth is imbued with power.

"Truth" thus has power in Foucault, but its power isn't what we're used to thinking about as power. First, no one forces us to believe in a truth; they simply teach us facts—things that they also accept as truth. Second, in Foucault, discursive power cannot be wielded by institutions, not even by states. Nor can discourses in Foucault's sense be deliberately controlled, not even by the seemingly powerful. Powerful institutions do work to create discourses (e.g., using public relations or advertising science), but they cannot make them catch on—they cannot give them power. Finally, "truth" can serve a resistance against states or other institutions with political-economic power. In Foucault's understanding, we cannot "speak truth to power." Our resisting truth is no more "true" than that of the politically and economically powerful.

Discourses accepted as true, powerful discourses, have the power to shape what we think and do, even in science and policy. People are not the source of discourses, even those that they themselves believe, and speak or write. But

Foucault's method offers us a resistance: if we analyze "truth" historically—or, I would add, ethnographically—we can come to question even what we ourselves believe. As Fairhead and Leach demonstrate in Chapter 3, truth regimes can be accessed analytically.

If knowledge has the power to govern by structuring the field of possibilities, then the knowledge accepted as true has the power to make alternative concepts and practices invisible. Truth governs "the things said and those concealed, the enunciations required and those forbidden." This is made explicit by James Ferguson in *The Anti-Politics Machine*, discussed in Chapter 3, in which the discourse of development makes politics disappear.

I am arguing that conservation is a discourse in a field of interrelated discourses, which shape how we know the world, what it is possible to think about the world, and how it is possible to practice in the world. But these discourses are not cohesive or stable. They are not constructed or controlled by any state or institution. We have all constructed them and imbued them with power together, through power relations. The elements of these discourses can be used, and reused, even by opposed players. Finally, these discourses can be analyzed by all of us, reducing their power over our thinking and practice. My hope is that these analyses, which we begin to examine in Chapter 3, can make a re-thought conservation more effective.

Notes

1. The interview question begins: "Does a mode of thought which introduces discontinuity and the constraints of system into the history of the mind not remove all basis for a progressive political intervention?" (Foucault 1991:53).
2. See also his lecture, delivered in 1969, entitled "What is an Author?" (Foucault 1977[1969]).
3. The second is the rule of continual variations: relations of power–knowledge are not static, but "matrices of transformations" (Foucault 1980:99). The third is the rule of double conditioning: no "local center" of power relations with their specific tactics could work for long without joining an overall strategy, and conversely such a strategy must have support from local power relations, "which serve as its prop and anchor point" (Foucault 1980:99). The two condition each other without being homogenous (Foucault 1980:100).

References

Foucault, Michel 1977[1969] What Is an Author? In *Language, Counter-Memory, Practice.* Donald F. Bouchard, ed. Pp. 113–38. Oxford: Basil Blackwell.

Foucault, Michel 1978[1976] *The History of Sexuality. An Introduction: Volume 1.* Robert Hurley, trans. New York: Vintage Books.

Foucault, Michel 1980 *Power/Knowledge: Selected Interviews and Other Writings 1972–1977.* Colin Gordon, ed. New York: Pantheon Books.

Foucault, Michel 1983 Afterword: The Subject and Power. In *Michel Foucault: Beyond Structuralism and Hermeneutics.* 2nd edition. Hubert L. Dreyfus and Paul Rabinow, eds. Pp. 208–26. Chicago: The University of Chicago Press.

Foucault, Michel 1991[1968] Politics and the Study of Discourse. In *The Foucault Effect: Studies in Governmentality with Two Lectures by and an Interview with Michel Foucault*. Graham Burchell, Colin Gordon, and Peter Miller, eds. Pp. 53–72. Chicago: The University of Chicago Press.

3

SEMINAL WORKS ON THE POWER OF DISCOURSES

Introduction

The second of Foucault's conceptual tools to be examined in this book is the power of discourses. James Ferguson, Arturo Escobar, and James Fairhead and Melissa Leach all produced books between 1990 and 1996 with enormous, even revolutionary impacts on the anthropology of development and conservation: Ferguson's *The Anti-Politics Machine* (1994a), Escobar's *Encountering Development* (1995), and Fairhead and Leach's *Misreading the African Landscape* (1996a). All of these authors were influenced by Michel Foucault's theories of discourses and their power. This means that Foucault's theories stimulated new ways of thinking—transformational new ways of thinking—about development and conservation.

The authors use a variety of terms for "discourse." Ferguson and Escobar actually use the term "discourse"; they also use the term "expertise," to refer to knowledge that is privileged ("truth" in Foucault). Escobar and Fairhead and Leach all use terms linked to perspective: visibilities, visions, gazes, and lenses. These terms are metaphorically rich: the power of a lens is that it shapes what is seen through it, as the "truth" in Foucault shapes what can be thought and practiced. Fairhead and Leach also use the words narrative and story, presaging Cronon in Chapter 4.

Though all three authors use Foucault's concept of powerful discourses, they can be contrasted in the way they do this. Ferguson makes the most use of Foucault's ideas, using multiple concepts. Escobar combines Foucault's concept of a discourse with power with the Marxist concept of domination. Fairhead and Leach are the least theoretical, but provide a full ethnographic elaboration of the history and effects of a powerful discourse.

All three of these seminal books can be read as ethnographies of conservation. Ferguson studied a mountain development project in Lesotho that focused on livestock and range management to stimulate a commercial livestock sector. Although economic development was clearly primary, range management was meant to be sustainable from the beginning, by focusing on improving forage and limiting overgrazing. The project failed to create a commercial livestock sector or change grazing patterns, because it misunderstood the role that livestock played in an economy that already employed most men in the area: that of the mine sector in South Africa. Escobar examined, particularly in Chapter 5, the implications of the discourse of sustainable development (new in the 1987 "Bruntland" report (World Commission on Environment and Development (1987)) for the environment in Colombia, particularly the focus on economic growth and managerialism. His study emphasizes the emergence of a discourse and its implications rather than the failure of a project. Fairhead and Leach carried out their research on forest islands in savanna in Guinea. Their book examines the discourses of pristine forest and locals causing degradation that led colonial and post-colonial scientists and policy-makers to misread the islands as vestiges of a forest rather than as locally created. They also look at the local story of the forest islands, and the daily practices that create them. They examine historical evidence to support that local story. Finally, they look at interactions between misguided forest conservation policies and local practices. The island forests are shaped by those interactions, not by either policy or local practice alone.

All of these authors carried out ethnographic fieldwork. James Ferguson carried out 15 months of field research in Lesotho (1982–3). Escobar carried out 15 months of field research in Colombia (1981–2 and 1983). Fairhead and Leach carried out a year of field research in Guinea (between 1992 and 1994).

The anti-politics machine

James Ferguson's[1] *The Anti-Politics Machine: "Development," Depoliticization, and Bureaucratic Power in Lesotho* launched the anthropology of development (versus the more applied development anthropology) in the United States. It also influenced anthropologists of conservation, because it analyzes a project in which grazing-land restoration depends on the adoption of modern livestock and grazing methods. The book was influenced by Foucault's *Discipline and Punish*, which Ferguson happened to take with him to Lesotho to carry out his dissertation research (Schouten 2009:2). The ideas in Ferguson's book that have been most influential were all influenced by Foucault: the "development machine," the demotion of intentionality, unintended political effects, and rendering these effects invisible (the anti-politics machine).

Ferguson uses the term "machine"—the "development machine," the "anti-politics machine"—borrowed from Foucault's *dispositif* or apparatus. The term has led, I think, to over-simplifications, but it is much more complex in Ferguson than sometimes remembered. I prefer the term "assemblage."[2] Development, which many see as a dominant discourse, with a machine-like coherence and purpose, actually doesn't dominate, and cannot be dominated. It is, rather, an assemblage of elements. To use Ferguson's (1994a:275) words, it is "an anonymous set of interrelations that only ends up having a kind of retrospective coherence," and "discursive systems link up with social institutions and processes without even approximately determining the form or defining the logic of the outcome." This logic has an "intelligibility" of its own, but "transcends the question of planners' intentions" (Ferguson 2006b:272–3). The "machine," combining discursive systems with social institutions and processes, and interacting with local social and cultural structures that are not acknowledged, appears to have coherence, logic, and intelligibility, but does not. The appearance of logic only emerges retrospectively, after the fact, and "counter-intentionally." In the 1990s this demotion of intentionality, which was very true to Foucault, was revolutionary. Power in conservation and development needed, and still needs, to be understood beyond powerful people or institutions or states and their purposes.

Ferguson used another idea of Foucault's, "instrument effects," to describe the logic of the outcome of the project, a logic that was not deliberately shaped or determined: "unintended effects." The idea of effects is related to the idea that there is no author: there is no intention or plan by those who appear to be in control (Ferguson 1994a:20). In Lesotho the unintended effects primarily bolstered "bureaucratic state power"; that is, the state of Lesotho ended up with a much stronger rural presence than before (roads, post offices, police, prisons, immigration control, army garrisons, etc., all understood in potentially negative terms), though this particular state, Lesotho, remained very weak. As Ferguson (1994a:273–4) puts it, "Specific bureaucratic knots of power are implanted," and "more power relations are referred through state channels." The state-strengthening results of development "emerge counter-intentionally through the working out of a complex and unacknowledged structure of knowledge in interaction with equally complex and unacknowledged local social and cultural structures" (Ferguson 2006b:281). These unauthored effects are clearly political, but Ferguson is careful not to overstate the actual power it gives to this weak state.

Another sort of unintended effect is much more important (as one can tell by the title of the book) and draws on Foucault's concept of discursive power: the development machine is an anti-politics machine, rendering politics invisible, making them disappear like the fictional anti-gravity machine makes gravity disappear (Ferguson 2006b:273). Development (at least in the 1980s) saw states as neutral and effective instruments for delivering economic growth. Even more damning, and fatal to the success of this particular project, the development machine erased Lesotho's position as a labor pool for South African mines. In the real political context, Lesotho laborers did not own cattle for profit, but to keep some of their wages from their relatives in a form that also retained their social position in their community. The "equally complex and unacknowledged local social and cultural structures," when combined with the erasure of political context, absolutely stymied project success. Before Ferguson, anthropologists did not have a clue how to understand such invisibilities. Ferguson's analysis led us to discursive power, and strengthened Foucault's ideas about the things unsaid, concealed, or forbidden.

Encountering development

Arturo Escobar[3] (1995:viii) read Ferguson's book while he was revising *Encountering Development*; he mentions Ferguson's "writings, discussions, and active support" in the book's preface. But he has his own direct links to Foucault from Berkeley, where he attended Foucault's seminar. The book's introduction cites an expansive literature (much of it dated after his dissertation), including postmodernists and post-colonialists as well as a great deal of Foucault's work. This huge literature contributes to making his writing style much looser than Ferguson's, but also impressive in its scope. Escobar is an enthusiastic and charismatic speaker and writer, and this enthusiasm did much to popularize Foucault in anthropology.

Encountering Development is intended to be an archeology of the discourse of development, examining "the deployment of the discourse through practices," the "concrete practices of thinking and acting through which the Third World is produced" (Escobar 1995:17, 11). Escobar's (1995:14) goal is "to contribute to the liberation of the discursive field so that the task of imagining alternatives can be commenced." He sees himself in this book as doing what Foucault did in *Discipline and Punish*.

Development is clearly seen by Escobar (1995:5) as a dominant discourse: "Certain representations become dominant and shape indelibly the ways in which reality is imagined and acted upon." The idea of a dominant discourse allows Escobar (1995:5–6) to "maintain the focus on domination," which he explicitly links to Marx. Domination exists in Foucault, but it is a fragile state of affairs, and discourses can both support and undermine it. In Escobar only peasant resistance movements can undermine the discourse of development. Still, it is clear that in reality development is a powerful discourse, and that it does shape the way many people see the world. Ferguson (1994a:xiii, 18) also sees it this way, saying that development is "a dominant problematic or interpretive grid through which the impoverished regions of the world are known to us," and "the thoughts and actions of 'development' bureaucrats are powerfully shaped by the world of acceptable statements and utterances within which they live."

Escobar (1995:162) also departs from Foucault, and from Ferguson in this case, in seeing the World Bank as "the master strategist" producing the development discourse that defines developing countries. As we saw in Chapter 2, Foucault does not see discourses as produced (they are authorless), and grand strategies, like discourses, emerge as local tactics converge and coalesce—which they do not always do. Escobar typically adds political economy to discursive power.[4] This is a laudable effort, but it leads him to over-simplify the emergence of the discourse of development. Escobar leaves unanswered the question of how the discourse of development emerged and gained power.

In Chapter 5 of *Encountering Development*, Escobar traces the "developmentalization" (in Colombia in the 1970s and 1980s, by the World Bank) of the small farmer (through integrated rural development), women (through women in development), and—most important for this book—the environment (through sustainable development). He argues that as new categories are brought into the development discourse, their subjects (farmers, women, the environment) are made visible in the largely economic terms of this discourse. His emphasis is on the discursive shaping of our vision, the "development gaze," a "regime of visuality": the World Bank learns about local societies through the "lens of neoclassical economics," or the "urban industrial vision" (Escobar 1995:162, 165). He bases this visual emphasis on Foucault's panopticon,[5] which produces a defining gaze architecturally, while the World Bank's gaze is discursive (Escobar 1995:155–6).

I concur with Escobar (1995:192–3) that the concept of sustainable development discursively healed the political conflict between environmentalists and developers, making us all believe not only that we could have development *and* conservation, but also that the best way to get to conservation in the developing world was through economic growth. The power of the discourse of development as economic growth is in fact demonstrated by its ability to absorb the environmental critique of economic development.

Escobar is the first thinker to bring Foucault and Marx together to understand the changing role of the environment. He links what he terms the economies of discourse (including the statements that development produces and the visibilities it organizes) and production (or political economy) (Escobar 1995:163). This is similar to Foucault (see Chapter 1), for whom the relation between power and the economy parallels that between power and knowledge. It anticipates Foucault's lectures on neoliberalism, lectures that would not have been available for Escobar to read. And Escobar takes this link between the economies of discourse and production and applies it to the environment. In his words, the current "reinvention of nature" is "effected by and within webs of meaning and production that link the discourses of science and capital" (Escobar 1995:211).

Misreading the African landscape

James Fairhead and Melissa Leach's[6] 1996 book, *Misreading the African Landscape*, is very influential in environmental anthropology. It is particularly valuable because it is a study of a socially created ecology, adding ethnography to ecology and vice versa (Fairhead and Leach 1996a:86).

Fairhead and Leach argue that what they call "a vision of degradation" was so strong that French colonial and contemporary Guinean policy-makers "read history backwards," interpreting forest islands in savanna as relics of an original forest destroyed by local land use, when in fact local land use created the islands out of savanna. Science, including social science, supported the false vision, which attests to its power. Fairhead and Leach oppose the degradation vision to the local vision—conservation discourse to local discourse—pitting their own research against the history of sciences that supported the conservation vision.

The vision of savanna as a degraded landscape, and the idea that local people must be responsible for that degradation, depends on the nature–culture conceptual dichotomy. The dichotomy associates original nature with primeval forest, not savanna. It also assumes that human society deforests, as though the conceptual distinction was written on the landscape. Fairhead and Leach document the local view of nature as not being opposed to the social: from the local point of view, forests historically shift in the landscape as settlement sites shift. For locals, the forests are "an archive of past habitation and sociality, as well as a landscape feature necessary for and shaped to meet present-day needs" (Fairhead and Leach 1996a:113).

The vision of forest degradation had very significant ramifications for conservation policy. For example, the social science version of this vision supported participatory policies, but the goal of participation was still the restoration of forests actually created by local people in the first place (Fairhead and Leach 1995a:85).

Fairhead and Leach also document the history of local practices responding to shifting policies informed by the conservation discourse, revealing a gap between policies and their on-the-ground effects. This undermines Escobar's sense of a discourse so dominant that it is able to alter local realities, and Ferguson's sense of a coherence of effects. In Fairhead and Leach a very strong and stable discourse just does not have much power to shape local realities. As they write, "Given the complexity of influences on these changes [in vegetation cover], it is hard to attribute a clear causal role to policy, especially policy whose implementation has been patchy and at times weak and ineffective" (Fairhead and Leach 1995a:83). Where environmental policy did have an impact, they argue "it was not because people adopted colonial suggestions wholesale, but because they used possibilities made available by policy" (Fairhead and Leach 1995a:84). They write that:

> resource management patterns have responded less to … policy than to changing social, economic and demographic pressures and possibilities. These have of course themselves been influenced by policies … but policies whose environmental effects are unforeseen … consequences.
>
> *(Fairhead and Leach 1995a:84)*

Finally, modern national policy still evinces multiple elements from the colonial degradation vision: shifting cultivation is irrational, fire is uncontrolled and dangerous, cattle in high numbers cause degradation, the population is ignorant, etc. (Fairhead and Leach 1991:84).

One example is the interaction between local fire practices and fire policies, which they trace from 1901 through the period of their fieldwork in the early 1990s. In both the dry uplands and the humid south of Guinea, they say, "villagers consider fire [timely fire-setting] as an integral part of the processes which shape their landscape" (Fairhead and Leach 1995a:62). They also understand the way that their farming and livestock practices influence fire patterns. One of the main local reasons for setting fires, for example, is to maintain pastures for livestock; they also observe that areas where livestock are put out to graze will not burn. Fire policies understand local fire practices as "setting bush fires," which has been criminalized since 1901. But fire policy and thus its criminalization has varied. The worst punishment was in the 1970s, when setting fires carried the death penalty (Fairhead and Leach 1996a:4). The most confusing policy to local people (instituted since 1935) was the setting of fires early in the dry season (Fairhead and Leach 1996a:35, 249). From the perspective of villagers, "Specific policies remained indistinct, within a general perception that forest guards banned villagers' fire-setting, and then sometimes (in early burning) set it themselves" (Fairhead and Leach 1996a:249). In the north, where setting fires was most necessary to villagers, they developed "coping and resistance strategies" like offering "hospitality and largesse" to forest officials, or setting fires secretly (Fairhead and Leach 1996a:253–4).

The effort in *Misreading the African Landscape* is to invest local understandings with truth value, or alternatively to give reality truth value. Fairhead and Leach (1996a) clearly believe in the power of discourses, and see a discourse as having warped science in the savanna. But they also clearly believe in the power of scientific evidence. As they say:

> Considering all landscape interpretations as in part socially constructed does not, however, negate the fact that certain readings can be demonstrated as false, and that historical evidence might support some more than others.
>
> *(Fairhead and Leach 1996a:3)*

"On the one hand we are dealing with landscape and its history as representation," they write, "but on the other hand we are attempting to reveal its empirical 'reality,' facts or events" (Fairhead and Leach 1996a:15–16). They see this as a methodological problem, and note the unease it causes. They meet this unease by widening their research to include an unusual array of sources: "archival documents, the writings of colonial anthropologists and Guinean scholars, policy documents and reports, old and new aerial photographs and maps, oral accounts, interview data in villages and policy circles, participant observation, and village resource and vegetation surveys"; along with "scientific works in ecology, forestry, botany, soil science, hydrology, and climatology" (Fairhead and Leach 1996a:16).

Fairhead and Leach (1996a:14) explicitly link the degradation vision to Foucault: "The conviction of degradation … and its power in excluding alternative landscape readings, may be considered as 'a discourse' in the Foucauldian sense." Their ethnography demonstrates just how far this tool can be taken in history and ethnography. They differ from Escobar in documenting the limits of discursive power, which influences the scientific and policy communities both before and after colonialism, but does not influence Guinean villagers—who cope with discourse-inflected policy as best they can.

Conclusion

These three works can be seen as ethnographies of the power that discourses can have. In Ferguson, Lesotho's position as a labor reserve for South Africa is ignored by organizations whose goal is the economic development of Lesotho. The discourse of development makes political context invisible, contributing to the failure of the project. In addition, in Ferguson's case study very real political effects, the extension of bureaucratic state power into the uplands, are rendered invisible. This influential case study underscores the power of discourses to make political context and political effects disappear; the discourse of development is, as Ferguson's book title says, an anti-politics machine.

In Escobar, the World Bank's discourse defines Colombia as a developing nation, "developmentalizing" small farmers, women, and nature with its economic "development gaze." Escobar stresses making things visible rather than invisible, but the discourse strips people and the environment of their own identities, histories, economies, and ecologies.

In Fairhead and Leach, Guinea's forest islands have been seen, since 1901, as vestiges of a primeval forest degraded by its inhabitants, rather than as a savanna in which the inhabitants have created forest islands. Conservation has read forest history backwards, interpreting a landscape through a discourse that assumes that forest is primeval and people degrade forest. This discourse lasted over a hundred years, inducing scientists, including social scientists, to support it. It shaped policy, and had serious effects on local inhabitants. In Fairhead and Leach's case study it is clearly a discourse of conservation that is sufficiently powerful to warp science and policy.

The power of discourses, however, is complex. In Ferguson the very real political effects of development are more accurately the unintended side effects of failed development. They can be seen and analyzed, but no amount of thoughtful planning can eliminate unintended effects. In Escobar, the discourses that define small farmer, women, and nature are strong, but historically changing. In Fairhead and Leach, the conservation discourse is strong, but its effects on the landscape are determined by complex interactions between policy, implementation, and local response.

All three of these authors make use of Foucault's conceptual tool of the power of discourses. Ferguson is the most accurate and uses multiple Foucauldian tools. Escobar is led astray to some extent by his early effort to combine Foucault's discursive power with Marxist dominance, particularly in attributing authorial power to the World Bank. This anticipates some of the problems that political ecologists will have over-simplifying the tool of discursive power. Fairhead and Leach provide the best ethnography of this tool, documenting how discourse-informed policy interacting with local practices was actually written on the landscape.

All three of these seminal ethnographies were based on long-term, field-based research, but they all also included research on the policy world. In Fairhead and Leach's case they cast their research net deep into history and wide into the policy world, as well as studying two villages in depth and others less deeply. This has become normal ethnographic research since the mid-1990s; it was much more unusual in the 1980s when Ferguson and Escobar were in Lesotho and Colombia. The research Fairhead and Leach carried out allowed them to document the history of the discourse, which Escobar also does, and to document the invisibilities and effects, as Ferguson does. But it also allowed them to document the process of policies (and science) developing alongside a discourse, and the uneven effects of policy on local Guineans.

In sum, the second tool drawn from Foucault, the power of discourses, stimulated three seminal works in the mid-1990s that have in their turn stimulated an anthropology of development and conservation that continues to grow. These three works demonstrate the usefulness of this tool. They also demonstrate the power of discourses, including the power of the conservation discourse, to warp policy and create project failure.

Notes

1. James Ferguson is professor and chair of anthropology at Stanford University. *The Anti-Politics Machine: "Development," Depoliticization, and Bureaucratic Power in Lesotho* was first published in 1990 (by Cambridge University Press). In 1994 it was re-published by the University of Minnesota Press. It was his revised dissertation. The best article-length summary of the book is Ferguson 2006b[1990]. For another summary, see Ferguson 1994b, which is the same as Ferguson 1997. Ferguson has published two other books: Ferguson 1999 and 2006a, along with many articles.
2. My use of "assemblage" follows Tania Li 2007. See also Chapter 15 in this book.
3. Arturo Escobar is an endowed professor of anthropology at the University of North Carolina at Chapel Hill. His most influential book is Escobar 1995, on development in the 1970s and 1980s in Colombia. Escobar was born in Colombia. He has an interdisciplinary PhD, *The Philosophy, Policy, and Planning of Third World Development (Anthropology, Political Economy, Critical Theory)*. He carried out 15 months of field research in Colombia (1981–2 and 1983). Like Ferguson's book, *Encountering Development* was a revised dissertation that became a groundbreaking text. Two other pieces, both published earlier than *Encountering Development*, also make Foucault's influence evident: Escobar 1984–5, a graduate paper of Escobar's developed in a Berkeley seminar with Michel Foucault in 1983; and Escobar 1988, an early version of Chapter 5 of *Encountering Development*. Escobar has also published Escobar 2008, and five co-edited volumes, along with many articles and chapters.
4. See also Escobar 1996 and 1999.
5. The panopticon appears in Foucault's *Discipline and Punish*.
6. James Fairhead and Melissa Leach are British anthropologists. He is professor of social anthropology and she is director of the Institute of Development Studies at the University of Sussex. They have written a lot about this case study, and a lot together. They carried out a year of field research in Guinea (between 1992 and 1994). They

met as doctoral students and were married in the middle of their fieldwork; they now have four children (Shephard 2007). In Fairhead and Leach 1996a they cite Foucault and Ferguson, but not Escobar (whose book was published just the year before). See also Fairhead and Leach 1995a, 1995b, 1996b, and 1997. They have also made a film on this fieldwork, *Second Nature*, dated 2006. They have written one other book together: Fairhead and Leach 2003.

References

Escobar, Arturo 1984–5 Discourse and Power in Development: Michel Foucault and the Relevance of his Work to the Third World. *Alternatives* X:377–400.

Escobar, Arturo 1988 Power and Visibility: Development and the Invention and Management of the Third World. *Cultural Anthropology* 3(4):428–43.

Escobar, Arturo 1995 *Encountering Development: The Making and Unmaking of the Third World.* Princeton: Princeton University Press.

Escobar, Arturo 1996 Constructing Nature: Elements for a Poststructural Political Ecology. In *Liberation Ecologies.* Richard Peet and Michael Watts, eds. Pp. 46–68. London: Routledge.

Escobar, Arturo 1999 After Nature: Steps to an Antiessentialist Political Ecology. *Current Anthropology* 40(1):1–30.

Escobar, Arturo 2008 *Territories of Difference: Place, Movements, Life, Redes.* Durham: Duke University Press.

Fairhead, James and Melissa Leach 1995a Reading Forest History Backwards: The Interaction of Policy and Local Land Use in Guinea's Forest-Savanna Mosaic, 1893–1993. *Environment and History* 1:55–91.

Fairhead, James and Melissa Leach 1995b False Forest History, Complicit Social Analysis: Rethinking Some West African Environmental Narratives. *World Development* 23(6):1023–35

Fairhead, James and Melissa Leach 1996a *Misreading the African Landscape: Society and Ecology in a Forest-Savanna Mosaic.* Cambridge: Cambridge University Press.

Fairhead, James and Melissa Leach 1996b. Enriching the Landscape: Social History and the Management of Transition Ecology in the Forest-Savanna Mosaic of the Republic of Guinea. *Africa* 66(1):14–36

Fairhead, James and Melissa Leach 1997 Webs of Power and the Construction of Environmental Policy Problems: Forest Loss in Guinea. In *Discourses of Development: Anthropological Perspectives.* R.D. Grillo and R.L. Stirrat, eds. Pp. 35–57. Oxford: Berg.

Fairhead, James and Melissa Leach 2003 *Science, Society and Power: Environmental Knowledge and Policy in West Africa and the Caribbean.* Cambridge: Cambridge University Press.

Ferguson, James 1994a[1990] *The Anti-Politics Machine: "Development," Depoliticization, and Bureaucratic Power in Lesotho.* Minneapolis: University of Minnesota Press.

Ferguson, James 1994b The Anti-Politics Machine: "Development" and Bureaucratic Power in Lesotho with Larry Lohmann. *Ecologist* 24(5):176–81.

Ferguson, James 1997 Development and Bureaucratic Power in Lesotho. In *The Post-Development Reader.* Majid Rahnema, ed. Pp. 223–33. London: Zed Books.

Ferguson, James 1999 *Expectations of Modernity: Myths and Meanings of Urban Life on the Zambian Copperbelt.* Berkeley: University of California Press.

Ferguson, James 2006a *Global Shadows: Africa in the Neoliberal World Order.* Durham: Duke University Press.

Ferguson, James 2006b The Anti-Politics Machine. In *The Anthropology of the State: A Reader.* A. Sharma and A. Gupta, eds. Blackwell Publishing. Pp. 270–86.

Foucault, Michel 1977[1975] *Discipline and Punish: The Birth of the Prison*. Alan Sheridan, trans. New York: Vintage Books.
Li, Tania 2007 Practices of Assemblage and Community Forest Management. *Economy and Society* 36(2):263–93.
Schouten, P. 2009 Theory Talk #34: James Ferguson on Modernity, Development, and Reading Foucault in Lesotho. *Theory Talks*, www.theory-talks.org/2009/11/theory-talk-34.html (22–11–2009).
Shephard, Jessica July 17, 2007 Melissa Leach: Village Voice. *The Guardian*, www.theguardian.com/education/2007/jul/17/highereducationprofile.academicexperts
World Commission on Environment and Development 1987 *Our Common Future: The Report of the World Commission on Environment and Development*. Oxford: Oxford University Press.

4

DISCOURSES OF CONSERVATION

Introduction

Foucault's concept of discourses and power, sometimes called post-structuralism, was not the only such idea at the time: other ideas influenced the social sciences and humanities in the 1980s and 1990s, and stimulated new critiques of conservation. These ideas were widespread, available in the works of symbolic anthropologists like Clifford Geertz; postmodern thinkers like James Clifford, George E. Marcus, and Michael Fischer; French philosophers like Michel Foucault, Jacques Derrida, and Bruno Latour;[1] American philosophers of science like Thomas Kuhn and Donna Haraway;[2] anthropologists like Gregory Bateson;[3] and Marxists like Raymond Williams and David Harvey. All of these writers, and more, informed the conceptual milieu that was brought to bear on conservation in the mid-1990s, culminating in the United States in Michael Soulé and Gary Lease's 1995 *Reinventing Nature?: Responses to Postmodern Deconstruction* and William Cronon's 1995 *Uncommon Ground: Rethinking the Human Place in Nature*, and in Great Britain in Roy Ellen and Katsuyoshi Fukui's 1996 edited book, *Redefining Nature: Ecology, Culture and Domestication*. These ideas were incredibly controversial,[4] though the controversy focused on Soulé and Lease and Cronon in America and not Ellen and Fukui in England. I will argue that these ideas remain potentially useful and have increased in importance. For some scientists and conservationists, unfortunately, the controversy continues to block the possibility of usefulness.

The controversy

Cronon

The controversy centered on the work of the environmental historian William Cronon.[5] His edited *Uncommon Ground: Rethinking the Human Place in Nature*, which includes his essay "The Trouble with Wilderness; or, Getting Back to the Wrong Nature," began in 1994 as a workshop at UC Irvine,[6] and was first published in 1995.[7]

The most significant intellectual inspiration for *Uncommon Ground*, and the "common ground" of its contributors, was not post-structuralism or postmodernism or deconstruction, but the cultural Marxist Raymond Williams' 1980 essay: "Ideas of Nature." As Cronon (1996:457–8) says in the book's conclusion:

> In thinking about a Big Message that I might offer as an ending for this book, I've kept returning to that phrase of Raymond Williams: "Ideas of nature, but these are the projected ideas of men." … It seems to me that all of us [the contributors] agree with this formulation of Williams, so our common ground … almost surely lies somewhere amid the unnaturally natural ideas and projections toward which he points.

In "The Trouble with Wilderness," Cronon (1996) traces the history of the American idea of wilderness to Western romanticism (nature is sublime) and the specifically American frontier myth. He argues that the central idea underlying "wilderness" is the separation between nature and people, such that getting back to nature is an individual quest, to find God or one's (masculine) power, alone, in nature. This conceptual separation, as Cronon (1996:80) writes, "leaves precisely nowhere for human beings actually to make their living from the land." Furthermore, he argues that the idea of wilderness makes environmentalists care about the wrong things: the "wrong nature" of the title. "Wilderness" as an idea erases "the history from which it sprang" (Cronon 1996:79):

> In its flight from history, in its siren song of escape, in its reproduction of the dangerous dualism that sets human beings outside of nature—in all of these ways, wilderness poses a serious threat to responsible environmentalism at the end of the twentieth century.
>
> *(Cronon 1996:81)*

Against this thread, "we need an environmental ethic that will tell us as much about *using* nature as about *not* using it" (Cronon 1996:83, author's italics). He softens this by saying that "deep reflection and respect must accompany each act of use, and … we must always consider the possibility of non-use" (Cronon 1996:89).

Environmentalists and conservationists

In the political context of a Republican-dominated Congress that opposed environmental protection, environmentalists and conservationists saw *Uncommon Ground* as an attack on the environmental movement and on conservation. This is clear in Soulé and Lease's 1995 volume *Reinventing Nature: Responses to Postmodern Deconstruction*, although that book is not specifically targeting Cronon.[8] The biologist Michael Soulé[9] writes that "postmodern deconstruction" constitutes a covert, ideological assault on nature and wilderness, that "serves to justify" an overt, physical assault by questioning the "existence and essential reality" of nature and wilderness (Soulé and Lease 1995:137). He details three postmodern myths: the myth of Western moral inferiority (aboriginal or Eastern attitudes towards nature are superior), the myth of constructionism (nature is unknowable or constituted by the economic activities of aboriginal peoples), and the myth of the pristine/profane dichotomy (the remnants of living nature are profane because they have been irreversibly changed by human disturbance) (Soulé and Lease 1995:146–7). The myth of constructionism questions "both the existence of nature and the accuracy of descriptions of it," and dethrones objectivism, such that biologists are no longer seen as gradually increasing knowledge of a real living nature (Soulé and Lease 1995:148). Soulé concludes:

> The nihilism and relativism of radically constructionist critiques of science and the materiality of nature, while popular in some academic circles, is sophomoric. Further, it is harmful because … it undermines efforts to save wildness and biodiversity.
>
> *(Soulé and Lease 1995:154)*

He is concerned that conservation policy is made by "bureaucrats, technocrats, planners, development specialists, lawyers, and economists" who are "trained by professors in the humanities and social sciences, many of whom are sympathetic to constructionist views" (Soulé and Lease 1995:161).

The environmental philosopher and deep ecologist George Sessions was one of the original critics of Cronon's book, writing that Cronon represents the "anthropocentric postmodernist approach." Sessions (1996:12) argues that Cronon's chapter "starts out by claiming that our concept of wilderness has to be re-thought or 'reinvented' in that it is a human or social construction. But by the end of the essay," he writes, "the tone changes significantly and the various human 'constructions' of wilderness … become largely irrelevant to the biological reality of protecting wild habitat." As we will see below, Cronon is arguing both that "wilderness" is a construction and that we need to protect the environment. Sessions (1996:12) assumes that postmodernists like Cronon believe "there is no reality beyond our human words and signs." In Cronon's case this assumption is wrong.

Sessions more recently demonstrated that the controversy is not over. He identifies William Cronon as a postmodern deconstructionist, saying that Cronon "claimed that wilderness protection poses a serious threat to environmentalism, which should be directed towards protecting the urban environment" (Sessions 2014:112). This is again a misunderstanding of Cronon, whose claim is rather that the *idea* of wilderness is a threat, because it limits conservation to those bits of nature perceived as wilderness.

Two environmental anthropologists have also recently renewed the attack on Cronon, Eleanor Shoreman-Ouimet and Helen Kopnina (2016:47), in their 2016 *Culture and Conservation: Beyond Anthropocentrism*. In their view, "Much of the contemporary interpretation of the cultural construction of nature has resulted from the work by William Cronon." They trace constructivism from French postmodernists (naming Lacan, Derrida, and Latour, but not Foucault) to political ecologists, naming Peet and Watts, Paul Robbins, Paige West, Arturo Escobar, and Peter Brosius, among others. Constructivism holds, in their view, the idea that "nature is wholly connected to the human perception of it": "Nature is not only represented by language but it is actually created by it," "this 'constructed' nature is completely detached from animals and plants and anything physical," "the conservation of nature is simply a manipulation of power over others, namely, poor developing countries," and "there is little 'real' or 'natural' about 'wilderness' outside of the human construction of it" (Shoreman-Ouimet and Kopnina 2016:45). These anthropocentric anthropologists, Shoreman-Ouimet and Kopnina (2016:49) write, "Deny the objective existence of nature outside of human perceptions of it and dismiss the gravity of its disappearance and destruction for humans, as well as plants, animals, and ecosystems." Furthermore, they write, "Arguments claiming that humans merely 'construct' nature … inevitably excuse the abuse of nature" (Shoreman-Ouimet and Kopnina 2016:58). Their argument echoes Sessions, and is equally unfair to and inaccurate about Cronon.

The political ecologist (and geographer) Paul Robbins defends political ecology from such attacks. He traces the idea of social construction to Foucault, and uses Fairhead and Leach (see Chapter 3) as a case study of its importance to political ecology. He makes a heroic effort to separate varieties of social construction, from "hard" or radical to "soft," which he says characterizes most political ecologists. Soft constructivism holds that "the objective world is real and independent of our categorizations but filtered through subjective conceptual systems and scientific methods that are socially conditioned" (Robbins 2012:128). I think this is exactly right. He mentions Cronon and the controversy (Robbins 2012:130). He argues, as I do, that "the political and ecological implications of this line of thinking [social construction] have proven useful in progressive research around the world" (Robbins 2012:131).

Cronon's defense

Cronon himself responds to the controversy in the new foreword to the 1996 paperback publication of the book. But the 1995 edition is already quite clear. In his section of the conclusion, he writes that the book was torn between two audiences:

> On the one hand, we need somehow to persuade scientists and environmentalists … that there is something profoundly important and useful in recognizing [nature's] cultural constructedness. On the other hand, we need no less to persuade humanists and postmodernists that … the world onto which we project those ideas is by no means entirely of our own making.… Nature is a mirror onto which we project our own ideas and values; but it is also a material reality that sets limits … on the possibilities of human ingenuity and storytelling.
>
> *(Cronon 1996:458)*

In the 1996 foreword, Cronon (21) argues: "Asserting that 'nature' is an idea is far from saying that it is only an idea.… And yet this is precisely the way some readers choose to interpret the message of this book."[10] He ends with the argument that the actions championed by environmentalists have behind them ideas that must be questioned, and that this questioning is crucial to environmentalism, and to "the human project of living on the earth in a responsible way" (Cronon 1996:22). Clearly the book failed to persuade scientists, some of whom believed that "postmodern deconstructionism" in general held the view that nature isn't real.

It is helpful to look back at a 1992 article written by Cronon, to see the development of ideas that were more fully expressed in the 1995 book. Cronon's "A Place for Stories," about two competing histories of the Dust Bowl, began to be written in 1987, and was published in 1992. This piece, and not "The Trouble with Wilderness," represents his first meeting with postmodernism. On one hand, in this piece he confronts the implications of two conflicting histories of the Dust Bowl for his own discipline, history. On the other, he confronts the implications of the "postmodernist assault on narrative" for his own commitment, as an environmental historian, to the narrative form (Cronon 1992:1348–9). He writes:

> For me, there is something profoundly unsatisfying and ultimately self-deluding about an endless postmodernist deconstruction of texts that fails to ground itself in history, in community, in politics, and finally in the moral problem of living on earth. Against it, I would assert the virtues of narrative as our best and most compelling tool for searching out meaning in a conflicted and contradictory world.
>
> *(Cronon 1992:1374)*

In 1992, Cronon (1373) would like to ground the "stories" of environmental history in nature: "Nature is hardly silent" and "Nature coauthors our stories." He sees the power of narrative to silence and erase but also to make us care: "A good story makes us *care* about its subject" (Cronon 1992:1374, author's italics). He ends by saying stories are unavoidable but must be used "consciously, responsibly, self-critically" (Cronon 1992:1376). One way to do this is to tell "stories about stories about nature," as he has done in this piece (Cronon 1992:1375).

Finally, in this piece Cronon confesses his own difficulties as an environmental historian confronting postmodernism. After spending five years writing four radically different versions of this essay, each version representing a different attempt to "acknowledge the immense power of narrative while still defending the past (and nature) as real things," he proposes another possible solution: histories cannot contravene known facts about the past, histories must make ecological sense, and we write as members of communities (Cronon 1992:1372–4). In 1992 Cronon was insisting that histories had to make ecological sense; by 1996 he had to defend his belief that ecology existed.

Cronon (1992:1350) states that a historical narrative "cannot avoid a covert exercise of power: it inevitably sanctions some voices while silencing others." Later on he says that powerful "hidden agendas" and "ideological needs"—which are Marxist understandings of power—influence the Foucauldian exclusions of discourse, in a way that is beyond the control of the author (Cronon 1992:1352). The competing narratives of the Dust Bowl turn out to be competing political narratives: anti–New Deal conservatism versus Worster's critique of capitalism and its failure to recognize natural limits (Cronon 1992:1362–3). Both histories together silence the Indian story of the Great Plains, which ends when they begin (Cronon 1992:1366).

I believe that Cronon's book was controversial because of the political climate in the United States when it was published: a Republican-dominated Congress was opposing environmental protection.[11] The political climate in America now is similarly bleak for conservation, and may be contributing to the renewal of the controversy.

A similar but uncontroversial book

Roy Ellen and Katsuyoshi Fukui's 1996 edited book, *Redefining Nature: Ecology, Culture and Domestication* can be read as the British and Japanese environmental anthropologists' version (though including several American anthropologists) of Cronon's American environmental historians' *Uncommon Ground*. It was based on a 1992 conference. In great contrast to Cronon's book, however, I can find no evidence that Ellen and Fukui's book excited any controversy. Yet the cultural construction of nature is one of the principle themes of the volume, because this idea "has become commonplace in anthropology and the history of ideas" (Ellen and Fukui 1996:3). Ellen, like Cronon, thinks about the difficulty of translating the idea that nature is constructed for natural scientists (Ellen and Fukui 1996:2).[12] "The meanings of nature are evidently myriad, multivalent and shifting, both between different populations and within them," he writes (Ellen and Fukui 1996:27). For anthropologists, this means that "we should not privilege our own concepts above those of our informants without first asking why we should do so and what the consequences are of so doing" (Ellen and Fukui 1996:28). But for biologists:

> The problematic and contingent—the fundamentally anti-monosemic—character of our constructions of nature is still not understood by many practitioners of science.... And the matter is an urgent and crucial one where science impinges on policy.
>
> *(Ellen and Fukui 1996:28)*

This includes conservation policy.

Ellen sometimes sounds like Cronon (1992), leery of the implications of post-modernism:

> We must be careful not to get caught up in a web of reflexiveness which will ultimately prevent us from explaining anything.... Recognition of infinite relativity ... is no basis for scientific comparison, intellectual communication or practical action.
>
> *(Ellen and Fukui 1996:28–9)*

Ellen concludes by saying that we need the concept of nature, much as Cronon (1992) says we need stories: "It is as if we cannot avoid a concept very much like nature to make sense of the world.... And this is not simply a folk compulsion, it is fundamental to the pursuit of science," by which he means biology and anthropology (Ellen and Fukui 1996:29). By his own logic, the concept of nature that we take forward in science should be one that recognizes that it is multiply constructed.

In a slightly later paper, Ellen uses just such a multiple and contingent concept of nature in a way that argues for the usefulness of this idea.

Usefulness of the idea of a discourse of conservation

Roy Ellen's[13] paper appeared in an edited volume (Li 1999) based on a 1995 conference at Dalhousie University in Canada; he was still working on his edited 1996 book during the conference. His paper gives us a rare ethnographic example of changing ideas about the forest; in essence, it tells us about the construction of a conservationist view of the forest.

Ellen (1999:139) writes, echoing his 1996 book, that "how people conceptualise nature depends on how they use it, how they transform it, and how, in so doing, they invest knowledge in different parts of it." But "such concepts are everywhere ambiguous, intrinsically moral in character and a *condition* of knowledge":

> Nature is not a *basic* category ... and means different—often contradictory—things in different contexts. It is constantly being reworked as people respond to new social and environmental situations.
>
> *(Ellen 1999:139, author's italics)*

His ethnography is about the Nuaulu people of Seram, Indonesia. Of the Nuaulu in particular, he writes:

> There is an evident underlying tension between an oppositional calculus of forests and "village" or "house," and a non-oppositional calculus which draws much more on the lived experience of particular strategies of subsistence which unite what we loosely call nature and culture.
>
> *(Ellen 1999:139)*

Nuaulu strategies of subsistence include the extraction of palm sago. The Nuaulu manipulate the vegetative reproduction of sago in a variety of ways, by selecting and protecting suckers, and by transferring root stocks closer to villages (Ellen 1999:137–8). Sago is ambiguously wild and domesticated (Ellen 1999:141). The Nuaulu also clear swiddens, but their agriculture is vegetative rather than seed based, meaning that it is based on tubers (Ellen 1999:140–1). These agricultural and ecological facts shape Nuaulu constructs, specifically the non-oppositional part of their concepts of nature and culture. We should also note that in order for Ellen to understand Nuaulu constructs, he had to first understand their agriculture, and the ecological character of sago palms and the ecological distinction between vegetative and seed-based propagation.

The Nuaulu have a word for forest, *wesie*, but it is—before the change—more anthropomorphic than oppositional to people (Ellen 1999:142). From the sixteenth century, however, when the Nuaulu are drawn into the global trade in spices and eventually the Dutch colony in the 1880s, a host of small changes—including the "decentering [of] sago from peoples [*sic*] conceptions of nature"—strengthens the opposition between *wesie* and people: owned land comes to oppose unowned *wesie*, and garden comes to oppose uncleared *wesie* (Ellen 1999:143–4). *Wesie* is now symbolically opposed to "house" (Ellen 1999:140). As Ellen (1999:144) explains, the links between *wesie* and subsistence were overshadowed by those between *wesie* and external exchange. Ellen (1999:144) argues that this movement from a non-oppositional to an oppositional model is "a dialectical function of a particular transitional history." Changes following the mid-1980s, especially commercial logging and transmigration, penetrated even more deeply, destroying the older logic based on a fundamentally limitless and implicitly sustainable forest. Logging was initially faced within the old logic: thus "an indigenous forest people appeared to be endorsing further forest destruction for short-term gain," because logging roads initially made reaching sago gardens easier (Ellen 1999:137). By 1990 the forest itself is seen as primarily a commodity, but it is a newly limited commodity, and one that must be defended, whether by negotiating with the state or being "quiet and obeying them" (Ellen 1999:147–8). The Nuaulu have become forest conservationists. In their own words: "We people find our food in the forest" (Ellen 1999:152).[14]

The Nuaulu, even the Nuaulu of the 1990s, use the forest—in Cronon's terms, they use nature. But there are uses and uses, and Ellen emphasizes two:

subsistence and external trade, or forest as food and forest as commodity. Forest as food—sago—does not allow a nature–culture dichotomy. And perhaps the very sustainability of this sort of subsistence doesn't allow the concept of a limited, must-be-defended nature. Forest as commodity—gradually, from spice to logs—makes its vulnerability clear in material terms as the Nuaulu are exposed to a "new level of discourse," in which "their" forest is part of a global economy that is also a "global ecology of limited goods" (Ellen 1999:150). The global economy altered local ecology in ways that were finally visible to and experienced by the Nuaulu, such that their construction of nature and their own identity was transformed.

There are many insights in this piece. Perhaps the most important is that ecology and constructions of nature are linked: they shape each other. The ecology of palm sago shaped its manipulation for subsistence that shaped local constructions of "forest." The logging of the forest re-shaped constructions of it, until the forest came to be seen as a limited commodity that must be defended. Environmental anthropologists need an understanding of ecology in order to understand local constructions of nature. Ecologists need to understand that local constructions of nature are shaped by local ecologies and the history of their use. Another important insight is that constructions of nature change as ecologies and their uses change. This has been and can continue to be studied. This is useful knowledge. It actually tells us a lot about how people become conservationists.

Conclusion

"Nature" and "wilderness" are discourses with power. That power can be seen in the controversy set off by Cronon's book. Though we might argue that the real enemy was never postmodernism nor constructionism/constructivism but Congress and presidents, and the political-economic power they have to disable conservation in the United States in order to "grow the economy." The power of these discourses can be seen in misreadings of Cronon and misunderstandings of political ecologists and environmental anthropologists. I've actually never met anyone who didn't believe in environmental reality, and I read a lot of Derrida and Foucault in graduate school, and a lot more since. I find the arguments of Soulé, Sessions, and Shoreman-Ouimet and Kopnina tangled and often spurious. I don't think many (if any) environmental anthropologists doubt environmental reality; Robbins says the same of political ecologists. Nor do I feel that politicians and policy-makers listen to environmental anthropologists or political ecologists.

I believe that people's perception and knowledge of ecology are strongly cultural. There is evidence for that in history, which Cronon pursued, and in ethnography, which Ellen carried out (and as we have seen in Chapter 3 that Fairhead and Leach did). The cultural character of "nature" and "wilderness" make it difficult for many environmental anthropologists to use the terms "nature" or "wilderness" without quotations; the quotations, however, mark the culture-shaped term, not any lack of reality.

Anthropocentrism is not opposed to ecocentrism. Most environmental anthropologists and political ecologists in particular do learn and use ecology; ecologists also need to learn that they themselves see and think about ecology through discourses of nature and wilderness, which are unlikely to be shared by local people in the global South. But for social scientists, ecologists, and local people, discourses of nature and wilderness are, as Ellen (1999:139) argues, "ambiguous, intrinsically moral in character and a *condition* of knowledge," mean "different—often contradictory—things in different contexts," and are "constantly being reworked as people respond to new social and environmental situations." This, it seems to me, is hopeful.

Notes

1. See also Chapter 7.
2. See also Chapter 17.
3. See Dove and Carpenter 2008.
4. These ideas are called "postmodern deconstructionism" (e.g., Sessions 2014), or simply constructionism, or alternately constructivism. The naming of these approaches is confusing, and seldom represents self-identification by authors. Foucault, for example, is called a post-structuralist, postmodernist, and social constructionist.
5. William Cronon is Frederick Jackson Turner Professor of History, Geography, and Environmental Studies at the University of Madison–Wisconsin. In 1992 Cronon left Yale, where he had taught since 1981 (while completing his Yale PhD in American history; tenured 1988), for Madison–Wisconsin.
6. The origins of Soulé and Lease's, and Cronon's, 1995 books both lie in a University of California Humanities Research Institute multi-year project (1992–4) at the University of California at Irvine, "Reinventing Nature," which was inspired by Donna Haraway's (1991) *Simians, Cyborgs, and Women: The Reinvention of Nature*, and included four conferences and a final residential workshop. Soulé and Lease's book came out of a (probably) 1992 conference at UC Santa Cruz, "Reinventing Nature?" Of interest is the fact that their discussions were facilitated by the postmodern anthropologist James Clifford. Cronon's book came out of the final residential workshop, held during the first half of 1994 at the University of California at Irvine, "Reinventing Nature."
7. *Uncommon Ground* came out in 1996 as a paperback, including a new foreword about the controversy surrounding its 1995 publication.
8. This book is commonly seen as a critical reaction to Cronon's book, but in fact neither Cronon nor his book are mentioned in it. As I clarify in Note 6, Soulé's conference preceded Cronon's workshop, and the books were published in the same year.
9. Soulé's concluding chapter does not mention William Cronon at all. He does mention Bruno Latour.
10. This reflects comments I heard at Forestry and Environmental Studies at Yale in 1997–8.
11. Another factor may be that Cronon's paper was "widely reprinted, in part, in major newspapers throughout the country" (Sessions 1996:12).
12. Donna Haraway contributed to Cronon's volume; Ellen finds her "persuasive" (Ellen and Fukui 1996:12).
13. Roy Ellen is a British anthropologist, until recently a professor of Anthropology and Human Ecology at the University of Kent.

14. Though these words are the Nuaulu representing themselves, more or less self-consciously, to outsiders.

References

Cronon, William 1992 A Place for Stories. *The Journal of American History* 78(4):1347–76.
Cronon, William, ed. 1996[1995] *Uncommon Ground: Rethinking the Human Place in Nature.* New York: W. W. Norton & Co.
Dove, Michael R. and Carol Carpenter, eds. 2008 *Environmental Anthropology: A Historical Reader.* Malden, MA: Blackwell.
Ellen, Roy 1999 Forest Knowledge, Forest Transformation: Political Contingency, Historical Ecology and the Renegotiation of Nature in Central Ceram. In *Transforming the Indonesian Uplands.* Tania M. Li, ed. Pp. 131–57. London: Routledge.
Ellen, Roy and Katsuyoshi Fukui, eds. 1996 *Redefining Nature: Ecology, Culture and Domestication.* Oxford: Berg.
Robbins, P. 2012. *Political Ecology: A Critical Introduction.* 2nd edition. London: Wiley-Blackwell.
Sessions, George 1996 Reinventing Nature, the End of Wilderness? A Response to William Cronon's Uncommon Ground. *The Trumpeter* 13(1):33–8.
Sessions, George 2014 Deep Ecology, New Conservation, and the Anthropocene Worldview. *The Trumpeter* 30(2):106–14.
Shoreman-Ouimet, Eleanor and Helen Kopnina 2016 *Culture and Conservation: Beyond Anthropocentrism.* London: Routledge.
Soulé, Michael and Gary Lease, eds. 1995 *Reinventing Nature?: Responses to Postmodern Deconstruction.* Washington, DC: Island Press.
Williams, Raymond, ed. 1980 Ideas of Nature. In *Problems in Materialism and Culture: Selected Essays.* Pp. 67–85. London: Verso.

5

THE TRIANGLE IN FOUCAULT

Introduction

Foucault's concept of governmentality is often applied to conservation: if you type "governmentality and conservation" into Google Scholar, you get about fourteen thousand results ("governmentality and environment" lists almost fifty-four thousand). This sizable literature argues that the concept of governmentality is relevant to conservation. The "governmentality lecture," given February 1, 1978, and first published in English in 1991, is typically cited when this concept is applied to conservation. I argue that governmentality is best understood in relation to the two linked concepts of sovereignty and discipline as they developed in Foucault's thought.

In Foucault, governmentality is a distinctive way of thinking about how to govern that emerges in the early nineteenth century, and is linked to the "modern" state and economy. Foucault contrasts it to sovereignty and discipline, ending the 1978 lecture by saying that sovereignty and discipline are not eliminated by governmentality; the three are a triangle (2007:102). I argue that in conservation interventions, all three types of government distinguished by Foucault are commonly found, sometimes in a single project. That triangle is the subject of this chapter.

The three types of government that constitute the triangle—sovereignty, discipline, and governmentality—provide three tools for our tool-box of ideas about the power of conservation.

I clarify two things in this chapter that make discipline and governmentality more applicable to conservation. First, I trace both ideas back to the concept of bio-power in *The History of Sexuality* (Foucault 1978), which allows me to apply insights developed in that book to discipline and governmentality. Second, I distinguish between discipline and governmentality. I argue that governmentality is often confused with discipline, to the detriment of an understanding of the subtle, indirect nature of governmentality.

When we implement conservation, we are governing people. The sort of government we are using is typically decided at the policy-making level. Foucault's concepts can be used by policy-makers to understand what type or types of government they are incorporating into their design, and by implementers to understand the government models implicit in the project they are implementing. Each type of government has different characteristics and implications for equity and for the effectiveness of conservation. The relation between types of government is of particular importance, as Chapter 6 will illustrate: sovereignty and governmentality, for example, do not combine well.

I should also make it clear that I do not condone any type of government. None of them promote equity. I believe that conservation would be most effective without violence, surveillance, or the manipulations typical of "participation." Socially aware conservationists need to work on imagining what conservation without government would look like. In the meantime, however, most policy-making and implementing conservationists must work within a system that governs, and understanding government is a place to start.

Foucault's sovereignty

Sovereignty is Foucault's term for the mode of governing in the feudal or medieval period, before the seventeenth or eighteenth centuries, in Europe. This coincides with a mercantilist global economy.[1] Sovereignty does not, however, end with the middle ages, as it continues even today in certain states and alongside discipline and/or governmentality in the triangle. Essentially, sovereignty for Foucault is the opposite of bio-power, or of both discipline and governmentality. In fact, it is the opposite of relations of power in Foucault's sense, as power (see Chapter 1) works through freedom or on free subjects. Sovereignty is thus a relatively sketchy concept in all of his work, used primarily to clarify contrasting concepts of power in which Foucault is more interested.

Sovereign power is "the right to decide life and death" (Foucault 1978:135). More precisely, Foucault (1978) explains, it is a power that can only be exercised where the sovereign's life is in jeopardy. When the sovereign is threatened by external enemies, he can wage war and require his subjects to defend him. "But if someone dared to rise up against him and transgress his laws, then he could exercise a direct power over the offender's life: as punishment, the latter would be put to death" (Foucault 1978:135). The right to decide life and death was "in reality the right to *take* life or *let* live" (Foucault 1978:135–6, author's italics). This power to take life existed in a historical context in which:

> power was exercised mainly as a means of deduction … a subtraction mechanism, a right to appropriate a portion of the wealth, a tax of

> products, goods and services, labor and blood, levied on the subjects. Power in this instance was essentially a right of seizure: of things, time, bodies, and ultimately life itself.
>
> *(Foucault 1978:136)*

Foucault provides three examples in which he contrasts sovereignty to discipline and governmentality. In the first, sovereignty responds to theft with a legal or juridical mechanism: a "completely simple penal law in the form of a prohibition" and punishment ("hanging, or banishment, or a fine") (Foucault 2007:4). This legal mechanism produces a binary division between what is permitted and what is prohibited and links types of prohibitions to types of punishments (Foucault 2007:5). In the second example, sovereignty responds to disease by exclusion. Lepers, for example, are simply not permitted to enter towns. Here the binary division is between those who are lepers and those who are not (Foucault 2007:9). In the third example, space and the town, sovereignty "is exercised within the borders of a territory," treating its citizens as a "multiplicity of subjects" or "a set of legal subjects" (Foucault 2007:11, 21). Walled towns were legally and administratively distinct from other parts of the territory, and more economically and socially mixed than areas outside the town (Foucault 2007:12). The capital should "capitalize" its territory: a prescriptive text from the mid-seventeenth century describes "a state well organized around a capital as the seat of sovereignty and the central point of political and commercial circulation" (Foucault 2007:15, 17). The relationship between town, sovereignty, and territory is imagined as the "interplay of macrocosm and microcosm" (Foucault 2007:16). In sum, sovereignty is characterized by simple laws that divide the permitted from the prohibited and link prohibitions to punishments, the exclusion of diseased persons from walled towns, and a territory politically, commercially, and symbolically centered on a capital as the seat of the sovereign.

Foucault (2007:89) also examines a literature on government, especially La Perrière's 1555 *Le Miroir Politique*, which was written against Machiavelli's *The Prince*. This sixteenth century literature describes the sovereign prince of *The Prince* as singular, external, and transcendent vis-à-vis his principality; because he has no connection to the principality, his relationship with it is "fragile and constantly under threat" (Foucault 2007:91). The Prince thus exercises power to strengthen and protect the principality—understood as the *relationship* of the Prince to his territory and subjects, the fragile link between them. In order to do this, the Prince analyzes threats and manipulates "relations of force" (Foucault 2007:92). In order to be a good sovereign, the Prince must consider "the common good," but the common good is essentially obedience or submission to the law: "The common good exists when all subjects obey the law without fail, perform their appointed tasks well, practice the trades to which they are assigned, and respect the established order" (Foucault 2007:98). Foucault (2007:98–9) notes that this is circular: "The good is obedience to the law, so that the good proposed by sovereignty is that people obey it"; sovereignty and law were "absolutely united." This returns Foucault (2007:110) to the law that was evident in the example of theft, and he characterizes sovereignty as "the state of justice, born in a feudal type of territoriality and broadly corresponding to a society of customary and written law."

In sum, sovereign power seizes wealth, labor, and potentially life itself. Simple laws and punishments typify sovereign power. Territory is politically and commercially oriented to the center of the Prince in the capital of the principality. The Prince is transcendent but threatened. Subjects are those who reside inside the territory, and they should obey the law, perform their labor, practice their trades, and respect the sovereign-headed order. Law and sovereignty are nearly synonymous, backed by relations of force, including the power to take life.

Sovereign power does occur in conservation. Parks are territorial units often governed by excluding some people; researchers and tourists, guards and guides may be allowed in, but villagers in buffer zones excluded, for example. This is very clearly sovereign power if violence or the threat of violence is the punishment for entering the territory, as sometimes happens.

Foucault's discipline

Discipline is Foucault's term for the mode of thinking about and practicing government from the mid-sixteenth century to the end of the eighteenth century in Europe, which he sometimes calls the Classical period. Like sovereignty, it coincides with a mercantilist economy. It brings a huge transformation in thinking about government: the difference between sovereignty and discipline is much more important than that between discipline and governmentality. Foucault's concept of discipline is also much more complex than his ideas about sovereignty. Much of *The Order of Things* (Foucault 1970) concerns knowledge in this historical period. *Discipline and Punish* (Foucault 1977) contrasts the punishments of sovereignty to disciplinary correction.

In *The History of Sexuality*, Foucault (1978) argues that mechanisms of power turn from the sovereign power to take life, to the administration of bodies and the management of life, generating, multiplying, and regulating life: this is bio-power. Bio-power has two halves, which are separate in the eighteenth century. The first half is concerned with bodies: disciplining the body "as a machine," "its disciplining, the optimization of its capabilities, the extortion of its forces, the parallel increase of its usefulness and its docility, its integration into systems of efficient and economic controls" (Foucault 1978:139).[2] This is what Foucault calls the anatomo-politics of the human body. The second half of bio-power is concerned with the population: how to control it, how to measure it, how to evaluate the relationship between resources and inhabitants, and how to adjust the accumulation of people to that of capital (Foucault 1978:140–1). This will become governmentality.

Foucault (2007) applies his three examples of theft, sickness, and space/towns to disciplinary power. Disciplinary power responds to theft with the disciplinary mechanisms of surveillance and correction, in contrast to the simpler law and punishment of sovereignty (Foucault 2007:4). Whereas sovereignty simply excludes lepers, discipline meets plagues with regulations that impose a "partitioning grid" on the town, with prohibitions and inspections controlling life even inside peoples' homes (Foucault 2007:10). In the third example, of space generally, sovereignty is exercised in a territory, but discipline "is exercised on the bodies of individuals" (Foucault 2007:11). Of towns, Foucault takes as his example artificial towns in France based on Roman camps, which are subdivided rectangles. These towns are also artificial constructions structuring an empty space, dealing only with the hierarchical and functional distribution of elements (Foucault 2007:16). This means that grids are laid down as though ecology did not exist. In contrast to the capitalized territory of sovereignty, discipline structures space in subdivided grids, which make surveillance possible during plagues. The grids of town planning resemble the grids of the great classificatory schemes of Classical knowledge in *The Order of Things*. Both are still with us.

Foucault (2007) also returns to his analysis of La Perrière's *Miroir Politique*, representing a sixteenth century literature written against Machiavelli and attempting to define an "art of government." This period is historically transitional between sovereignty and governmentality. In this transitional period, "government" first emerges as a problem and as the state's rationality; the object of government, however, is the state, no longer the sovereign, but not yet the population. The art of governing the state is like the art of governing the family: the government of the family is called "economy," and it involves exercising "supervision and control over its inhabitants, wealth, and the conduct of all and each" (Foucault 2007:94–5).

Governing in the disciplinary period is "the right disposition of things arranged so as to lead to a suitable end" (Foucault 2007:96, quoting La Perrière). "Things" is actually a "complex" of men and things:

> men in their relationships, bonds, and complex involvements with things like wealth, resources, means of subsistence, and, of course, the territory with its borders, qualities, climate, dryness, fertility, and so on. "Things" are men in their relationships with things like customs, habits, ways of acting and thinking. Finally, they are men in their relationships with things like accidents, misfortunes, famine, epidemics, and death.
>
> *(Foucault 2007:96)*

La Perrière's "ends" are multiple and specific to the things being governed. The good governor, quoting La Perrière again, has "patience, wisdom, and diligence": he reigns without needing a sting, with knowledge of things and their right disposition, and as if he served his subjects (Foucault 2007:99–100). Foucault (2007:100–1) takes care to note that the art of government existed in reality, not just books: the administrative apparatus is developed, along with new forms of knowledge, especially statistics.[3]

Disposition means the way something is arranged or ordered in relation to other things. The "right disposition" of the art of government reflects the great classificatory schemes of Classical knowledge in *The Order of Things.* In the examples the partitioning grid of response to the plague and the subdivided rectangles of the new towns allow for "right disposition" of people. All are imagined as a grid or table, and disposing is imagined as finding the right box for a newly discovered plant or a particular category of people.

In this same historical period, discipline and surveillance emerge in institutions like prisons, schools, factories, and armies, aided by new architectural structures like the panopticon,[4] which arrange inmates, pupils, workers, and soldiers so they are all visible to surveillance. In prisons, schools, factories, and armies, "the conduct of all and each" can be supervised and controlled, as a father of his family. These institutions and their buildings continue to be disciplinary today.

The art of government seeks perfect control of all and each even outside such institutions. Government seeks "perfection, maximization, or intensification of the processes it directs" (Foucault 2007:99). This is the "great disciplinary dream" (Foucault 2007:341). It leads to efforts to regulate every imaginable aspect of a perceived problem. In order to prevent grain scarcity in seventeenth and eighteenth century France, for example, prices were controlled, hoarding was prohibited, exports were limited, the amount of land that could be cultivated was limited, but minimums were also set, growing some non-grain crops was prohibited, merchants were forced to sell stocks immediately, stocks were checked at first harvest, and transport of grain by ship was prohibited (Foucault 2007:31–2). All of this required multiple regulations and a massive bureaucracy to administer and enforce those regulations. The need for statistics is clear. Note that these disciplinary regulations are not the simple law of sovereignty; they are law as tactic, law used to achieve an end, the solution to a problem.

I argue that this art of government period is primarily disciplinary in Foucault's thought. The historical texts on government that he follows trace a series of changes from advice to the Prince on how to hold on to the principality (Machiavelli), to treatises on the art of government (La Perrière), and finally to political science; from the state of justice, to the administrative state that "corresponds to a society of regulation and disciplines," to the gradually "governmentalized" state (Foucault 2007:88, 108–9, 110). Governmentality was unable to emerge as long as it was trapped between the "excessively large, abstract, and rigid framework of sovereignty" and the art of government's "too narrow, weak, and insubstantial" model of the family (Foucault 2007:102–3).

In sum, disciplinary power governs life itself. It administers bodies, disciplining them and disposing them in grid-like, partitioned spaces, as towns and in prisons, schools, factories, and armies. It governs a complex of people and things, seeking to perfect, maximize, or intensify political and economic processes. This is associated with a large administrative state, which produces disciplinary regulations and undertakes surveillance and correction. The new knowledge of statistics assists the state. The state's primary model is the family: the state must be governed as a father governs his family, with meticulous attention. Disciplinary power dreams of perfection: of every thing and every person being in place, in order.

Like sovereignty, discipline is still very much alive; wherever we see surveillance, grid-like spaces, and lots of regulations and enforcement, we are seeing discipline. If the model for organizing a forest guard, for example, is a military model, the guards are being governed in a disciplinary way and probably enforcing in a disciplinary way too. If trainings take on attributes of a school they may also be suspect. I believe that conservation in particular has a tendency towards disciplinary government, because conservationists tend to imagine local people as a threat to forests or wildlife, as out of place and out of control. Conservation tends to govern people to maximize forests or wildlife, and that alone is already disciplinary.

Foucault's governmentality

The fourth point I made about power in general, in Chapter 1, is that for Foucault power is always associated with freedom. This is particularly true of governmental power, which is the sort of government practiced by modern, liberal states. I want to begin talking about Foucault's concept of power in general, and then move to governmentality. Foucault (1983:220) writes that "what defines a relationship of power is that it is a mode of action which does not act directly and immediately on others. Instead it acts upon their actions." The exercise of power, he continues, "is a total structure of actions brought to bear upon possible actions; it incites, it induces, it seduces, it makes easier or more difficult" (Foucault 1983:220). Foucault (1983:220–1) then introduces the term "conduct," "for to 'conduct' is at the same time to 'lead' others … and a way of behaving within a more or less open field of possibilities." This is the source of the common definition of governmentality as the conduct of conduct. Foucault (1983:221) continues, saying that "basically power is less a confrontation between two adversaries … than a question of government," when the term government "designated the way in which the conduct of individuals or groups might be directed: the government of children, of souls, of communities, of families, of the sick.… To govern, in this sense, is to structure the possible field of action of others." To govern in this way is governmentality.

The term governmentality is first coined by Foucault in a lecture given early in 1978. But important concepts exist earlier, especially regulatory controls, regulation, calculated management, or apparatus of security. In Foucault 2007 (24f5), the editor tells us that Foucault first distinguishes security mechanisms from disciplinary mechanisms in the final lecture of 1975–6 (Foucault 2003:246), but does not use the term in *The History of Sexuality* (1978[1976]). In that work Foucault (1978:145) opposes discipline, which is exercised on the bodies of individuals, to "'regulatory controls' that take charge of the health and life of populations."

Governmentality characterizes the mode of government that follows the art of government transitional period, not replacing but subsuming discipline early in the nineteenth century. The economy in Europe is transitioning to what will be called capitalism. Governmentality is clearly distinct from sovereignty, and I argue that it is also different from discipline.

Foucault (1978) writes that early in the nineteenth century the two halves of bio-power, bodies and population, which were separate in the eighteenth century, come together. The first is the anatomo-politics or administration of the human body. The second will become governmentality focused on the population (in the statistical sense of the word), but in the eighteenth century it focused on the "species body," "the body imbued with the mechanics of life and serving as the basis of the biological processes: propagation, births and mortality, the level of health, life expectancy and longevity" (Foucault 1978:139). The species body was governed by "an entire series of interventions and *regulatory controls: a bio-politics of the population*," which he also calls "the calculated management of life" (1978: 139–40, author's italics). The anatamo-politics of the body and the bio-politics of the population, then, come together in the nineteenth century. They come together not discursively but through a series of "concrete arrangements," for example, sexuality, the subject of Foucault's history. The politics of sexuality joins body and population, discipline and regulation (Foucault 1978:139).

Foucault links this joining of bodies and population explicitly to the development of capitalism. Capitalism would not have been possible, he writes, "without the controlled insertion of bodies into the machinery of production and the adjustment of the phenomena of population to economic processes" (Foucault 1978:141). The essential thing was to optimize "forces, aptitudes, and life in general without at the same time making them more difficult to govern" (Foucault 1978:141).

We can return to Foucault's (2007) three examples. The first example is theft. In the nineteenth century, law, punishments, surveillance, and correction remain unchanged, but begin to be informed by an entirely different sort of question:

> What is the average rate of criminality for this [type, i.e., theft]? How can we predict statistically the number of thefts at a given moment, in a given society, in a given town … in a given social stratum, and so on? … are there times, regions, and penal systems that will increase or reduce this average rate? … What is the cost of repressing these thefts? … What … is the comparative cost of the theft and of its repressions.… The general question basically will be how to keep a type of criminality, theft for instance, within socially and economically acceptable limits and around an average that will be considered optimal for a given social functioning.
>
> *(Foucault 2007:4–5)*

Governmentality "inserts [theft] within a series of probable events," then inserts it in a "calculation of cost," then "establishes an average considered optimal on the one hand, and, on the other, a bandwidth of the acceptable that must not be exceeded" (Foucault 2007:6).

The same calculations govern the second example, inoculation:

> The fundamental problem will not be the imposition of discipline, although discipline may be called on to help, so much as the problem of knowing how many people are infected with smallpox, at what age, with what effects, with what mortality rate … the risks of inoculation, the probability of an individual dying or being infected … despite inoculation, and the statistical effects on the population in general.
>
> *(Foucault 2007:10)*

We are all familiar with these problems, which continue to surround flu epidemics. Foucault also says that medical campaigns now try to stop such epidemics.

For the third example, Foucault (2007:17) uses the real development of the existing town of Nantes, planned in order to relieve overcrowding, make room for new economic and administrative functions, improve links to the hinterland, and allow for growth. The transformation of Nantes involved "cutting routes through the town, and streets wide enough to ensure four functions": hygiene, trade, connections to external roads, and surveillance (Foucault 2007:18). "It was a matter of organizing circulation … making a division between good and bad circulation, and maximizing the good circulation by diminishing the bad" (Foucault 2007:18). In addition to this, there was the "new and fundamental question of how to integrate possible future developments within a present plan" (Foucault 2007:18). Foucault (2007:19–20) lays out the new elements in this example: space is no longer imagined as empty (security considers water, islands, air, etc.; that is, ecology); the goal is the maximization of the positive elements, not their perfection; elements are poly-functional; and plans must account for the future. He calls this space the "milieu." He selects this term, borrowed from physics, because it is "the medium of an action and the element in which it circulates," addressing "the problem of circulation and causality": "It is what is needed to account for action at a distance"; it is "a field of intervention" (Foucault 2007:20–1). I think the milieu is the space in which one imagines conducting the conduct of a population.

Historically, during the transitional period (which I have linked to discipline) of the seventeenth and early eighteenth centuries, the art of government was blocked from becoming governmentality. Its unblocking "was linked to the emergence of the problem of the population," alongside the emergence of the science of government and the re-focusing of the economy away from the family (Foucault 2007:104). Foucault (2007:108–9) defines governmentality as three things: "the ensemble formed by institutions, procedures, analyses and reflections, calculations, and tactics that allow the exercise of … power that has the population as its target, political economy as its major form of knowledge, and apparatuses of security as its essential technical instrument"; the preeminence over sovereignty and discipline of the type of power that we can call "government"; and, finally, the "governmentalization" of the state. I will further examine the population, political economy, and security.

The population becomes the "final end of government": "to improve the condition of the population, to increase its wealth, its longevity, and its health" (Foucault 2007:105). The improvement of the population does not mean that the population is consulted about what it wants; on the contrary, it is "the object of government manipulation; vis-à-vis government, [population] is both aware of what it wants and unaware of what is being done to it" (Foucault 2007:105). The welfare of the population as the goal of government replaces the state itself in the disciplinary period and the sovereign under sovereignty.[5]

The new science of government was first called "political economics" but came to be called economics (Foucault 2007:106). Political economy "is absolutely inseparable from the constitution of a knowledge of the processes revolving around population in the wider sense of what we now call 'the economy'" (Foucault 2007:106–7). In the seventeenth century, "economy" was reduced to the family model. By the eighteenth century, "the economy" "will designate a level of reality and a field of intervention for the government" (Foucault 2007:95). And the discipline of economics has come to be the core knowledge of governmentality.

Both the population and the economy were understood as "subject to natural processes," which meant that the state's role became "to respect these natural processes," "to arouse, to facilitate, and to *laisser faire*, in other words to manage and no longer to control through rules and regulations" (Foucault 2007:352–3, author's italics). The liberal state had to respect the freedom and autonomy of the population and the economy.[6]

However, this freedom made apparatuses[7] of security necessary. Liberalism affirms the necessarily autonomous and opaque character of the processes of population and economy, but it is preoccupied with the vulnerability of these processes, which thus need to be framed in mechanisms of security. Mechanisms of security ensure "the security of the natural phenomena of economic processes or processes intrinsic to population" (Foucault 2007:353). They also affirm the vulnerability of these processes, and thus the need for interventions (Osborne 1996:102). So, the freedom and autonomy of the population and the economy are affirmed but also constitute dangers to liberal government. The freedom of the individual is also a danger. Foucault (2008:65) says that liberalism calculates the point at which individual interests constitute a danger for the population—this is the problem of security—but conversely individual interests must not be encroached upon—this is the problem of freedom. Security is "both liberalism's other face and its very condition," and "the game of freedom and security is at the very heart of this new governmental reason" (Foucault 2008:65).[8]

What does security look like? It is management and planning; using statistics and the emerging discipline of economics to conduct the conduct of the economy, the population, and individuals, in order to ensure the autonomy of the economy; the welfare and autonomy of the population; and the freedom of individuals. Foucault gives the historical example of the physiocrats[9] trying to imagine an apparatus:

> for arranging things so that, by connecting up with the very reality of these fluctuations [between grain abundance/scarcity and dearness/cheapness], and by establishing a series of connections with other elements of reality, the phenomenon is gradually compensated for, checked, finally limited, and, in the final degree, canceled out, without it being prevented or losing any of its reality.
>
> *(Foucault 2007:37)*

The idea is to work within the fluctuations in supply and price, and not try to prevent them in advance, as disciplinary government did. Discipline regulated everything; the apparatus of security, in contrast, "lets things happen" (Foucault 2007:45).[10] The avoidance of governing too much superseded the disciplinary avoidance of governing too lightly.

The freedom of the population, the economy, and individuals characteristic of governmentality makes governing incredibly complex. Foucault (2007:20), in his example of the milieu, writes that the problem of security is essentially the problem of an "indefinite series of mobile elements" (e.g., carts, thieves, miasmas), an "indefinite series of events" (e.g., carts will arrive), and an "indefinite series of accumulating units" (inhabitants, houses) (Foucault 2007:20). Security, he continues:

> will try to plan a milieu in terms of events or series of events or possible elements, of series that will have to be regulated within a multivalent and transformable framework. The specific space of security refers then to a series of possible events; it refers to the temporal and the uncertain which have to be inserted within a given space.
>
> *(Foucault 2007:20)*

These series are open or indefinite, so they "can only be controlled by an estimate of probabilities" using statistics. He concludes that the management of these series "is pretty much the essential characteristic of the mechanism of security" (Foucault 2007:20). Freedom and security are at the heart of governmentality.

To attempt to sum up governmentality:

1 Governmentality is conducting or directing conduct indirectly, by structuring the possible field of action of others.
2 Those whose actions are conducted by governmentality are free to act. Governmentality is liberal government.
3 The improvement of the welfare of the population is the goal of governmentality, but the population is also the object of governmental intervention,

unaware of what is being done to it. It is imagined as a natural phenomenon, whose autonomy or freedom must be respected. This freedom is perceived as a danger.

4 The economy is also imagined as a natural, autonomous, free phenomenon. Its freedom is also perceived as a danger.

5 The contradiction between individual freedom and the population is another danger that freedom poses for governmentality.

6 The processes of population and economy are complex: managing them involves planning for "open series" of multiple variables moving in space and time. This sort of planning depends on statistics, which allow for estimates of probabilities and for balancing the relative risks and costs of alternatives.

7 This is what "mechanisms of security" are: management and planning to protect autonomous, complex processes and the contradiction between individuals and the population.

8 Political economics and then economics are the knowledges of governmentality.

9 Governmentality tries not to govern too much; it tries to let things be, to *laisser faire.*

Conservation policies and interventions involving local people that do not depend on either violence or discipline (multiple regulations and surveillance) are probably all "governmental" in Foucault's sense. The good thing about this is that the rationality is a positive one: improving people's welfare. But the intent is still to change behavior, to "conduct their conduct." The assumption behind this intent is that local people need changing—improving interventions are necessary. In conservation this is joined to the discourses that oppose people to nature and suspect people of degrading wilderness and threatening wildlife. This model of governing alongside these discourses prevents necessary research. How do local people now and historically interact with the environment around them? Do they perceive the threats conservationists see? If so, how are they responding to these perceptions?

In addition, the subjects to be improved in governmentality are not directly told what is being done to them, and this is true in conservation too: conservation interventions whose main goal is reforestation, for example, may be presented to local populations as opportunities to profit from nearby forests. Governmentality acts on people—it doesn't sit down with them and directly share goals and negotiate compromises.

Finally, conservation and the organizations doing conservation all manage and plan; this has become simply the way things are done in modern government, but it is government. It determines our stance towards the environment as well as the people using it. When conservationists design or implement a policy they are attempting to manage people, which means to govern them. There is power in this stance. People are complex to manage and perceptive about power.

Conservationists increasingly have degrees in governmentality's knowledges: economics, management, and statistics. These knowledges have the power to prevent conservationists from seeing the power and government inherent in conservation. Ethnographers, as Tania Li (2007:9) says in an essay on governmentality, "have work to do."

Conservation projects that intend to transform local people into conservationists are also governmental, but this topic will be treated separately, in Chapter 9.

Foucault's triangle

Foucault's often-cited lecture on governmentality, as we have seen above, presents sovereignty, discipline, and governmentality historically, as following one another. But at the end of that lecture Foucault (2007:107–8) corrects this:

> We should not see things as the replacement of a society of sovereignty by a society of discipline, and then of a society of discipline by a society, say, of government. In fact we have a triangle: sovereignty, discipline, and governmental management.

By this he means that the three sorts of government are all present at once, both in the past and today.

To the best of my knowledge, the triangle does not appear in Foucault's (2007:6) writing until the first lecture of 1978 (the governmentality lecture is the fourth), when Foucault says that the "ancient modalities" of power "involve those that appear as newer." Thus, in sovereignty, which "was dominant" until the eighteenth century, "the disciplinary side was far from being absent": exemplary punishment (public torture and execution), for example, was supposed to have a corrective, disciplinary effect on its witnesses (Foucault 2007:6). Governmentality was also present in sovereignty: crimes by servants were severely punished due to their probability, a mechanism of security (Foucault 2007:7). Similarly, discipline included governmentality: the correction of prisoners is gauged according to their risk of recidivism, which is a mechanism of security (Foucault 2007:7).

Looking forward historically, as mechanisms of security (governmentality) develop, there is no "bracketing off or cancellation of juridico-legal structures or disciplinary mechanisms." Still drawing examples from the prison system, Foucault (2007:7) says that "getting these systems of security to work involves a real inflation of the juridico-legal code" of sovereignty. Similarly, "There is a considerable activation and propagation of the disciplinary corpus" (Foucault 2007:7). Sovereignty blocked the development of governmentality: "So long as sovereignty was the major problem and the institutions of sovereignty were the fundamental institutions … the art of government could not develop" into governmentality (Foucault 2007:102). Yet the problem of sovereignty "was never more sharply posed" than when the art of government became governmentality" (Foucault 2007:106). Governmentality retains discipline also: all the disciplinary institutions—schools, workshops, armies—continue. The management of the population, which is governmental insofar as it is managed as a "collective mass of phenomena" or at the level of "overall results," was also disciplinary insofar as it was managed "in depth, in all its fine points and details" (Foucault 2007:107).

Historically, one mode of government at a time may "dominate" or, perhaps, be the "essential mechanism." But each mode, even speaking historically, included the other two in some way. This means that we must not look for governmentality alone in contemporary conservation, or even for governmentality and discipline together; we need to look for the triangle. We need to ask which if any mode of government dominates or is essential, and how functions are divided between the three.

Conclusion

In sum, Foucault differentiates between the three models of government that make up the triangle. We can analytically separate the sorts of government Foucault describes, but in reality they all occur together. These three models provide three tools for our tool-box of ideas about power in conservation.

The biggest difference is between sovereignty on the one hand and discipline and governmentality—which together constitute bio-power—on the other. Sovereignty does not govern life; discipline and governmentality do. With bio-power, Foucault (1976:142–3) writes, life "passed into knowledge's field of control and power's sphere of intervention." Bio-power governs life at two levels: the body and the population. I have argued that discipline is the governing of the body, and governmentality the governing of the population.

I have also argued that discipline should not be confused with governmentality. The two together govern life, not only at different scales, but in very distinct ways. Disciplinary power dreams of a perfect, grid-like structure, which provides a box for everything and everyone. This leads to disciplinary regulations and surveillance, and a huge administrative state. Governmentality thinks quite differently, of improvement and optimization, rather than perfection. Governmentality tries not to govern too much. But mechanisms of security are necessary to manage and plan for—to both protect and cope with—the freedom of the population and the economy.

As I noted above, I do not condone any type of government. I believe that conservation would be most effective without violence, surveillance, or the manipulations typical of "participation." I believe that conservationists need to begin imagining what conservation without government would look like.

Finally, I have argued that all three modes of government occur in conservation. Chapter 6 will provide four ethnographic examples.

Notes

1. *Mercantilism* was an economic theory and practice in Europe from the sixteenth to the eighteenth century that "promoted governmental regulation of a nation's economy for the purpose of augmenting state power at the expense of rival national powers" (The Editors of Encyclopaedia Britannica 2019).
2. See also "Docile Bodies" (Foucault 1977:135–69).
3. Statistics means science of the state.
4. See "Panopticism" (Foucault 1977:195–230).
5. Foucault (2007:43) notes that the physiocrats (see Note 9) said that people in the population should conduct themselves in a certain way (e.g., sometimes accepting high grain prices or grain scarcity): some people might put themselves outside of the population and thus disrupt the system by either revolting and seizing supplies or by monopolizing supplies.
6. This is also essential to the development of capitalism ("*laisser faire, passer et aller*") (Foucault 2007:48).
7. Foucault intends apparatus simply as a heterogeneous ensemble or assemblage. See Chapter 15.
8. See also Osborne 1996:117.
9. The physiocrats were eighteenth-century French economists. See Chapter 11.
10. See Chapter 11.

References

The Editors of Encyclopaedia Britannica 26 June 2019. Mercantalism. *Encyclopaedia Britannica.* https://Britannica.com/topic/mercantilism. accessed 11/17/2019.

Foucault, Michel 1970[1966] *The Order of Things: An Archaeology of the Human Sciences.* New York: Vintage Books.

Foucault, Michel 1977[1975] *Discipline and Punish: The Birth of the Prison.* Alan Sheridan, trans. New York: Vintage Books.

Foucault, Michel 1978[1976] *The History of Sexuality: Volume I: An Introduction.* R. Hurley, trans. New York: Vintage Books.

Foucault, Michel 1991[1978] Governmentality. In *The Foucault Effect: Studies in Governmentality.* Graham Burchell, Colin Gordon, and Peter Miller, eds. Pp. 87–104. Chicago: University of Chicago Press.

Foucault, Michel 2003 *"Society Must Be Defended": Lectures at the Collège de France 1975–1976.* Mauro Bertani and Alessandro Fontana, ed. David Macey, trans. New York: Picador.

Foucault, Michel 2007 *Michel Foucault: Security, Territory, Population: Lectures at the College de France, 1977–1978.* Michel Senellart, ed. Graham Burchell, trans. New York: Picador.

Foucault, Michel 2008 *The Birth of Biopolitics: Lectures at the College de France, 1978–1979.* Michel Senellart, ed. Graham Burchell, trans. New York: Picador.

Li, Tania M. 2007 Governmentality. *Anthropologica* 49(2): 275–81.

Osborne, Thomas 1996 Security and Vitality: Drains, Liberalism and Power in the Nineteenth Century. In *Foucault and Political Reason: Liberalism, Neo-liberalism and Rationalities of Government.* Andrew Barry, Thomas Osborne and Nikolas Rose, eds. Pp. 99–122. Chicago: The University of Chicago Press.

6

SOVEREIGNTY, DISCIPLINE, AND GOVERNMENTALITY IN ETHNOGRAPHIES

Introduction

In this chapter I will examine sovereignty, discipline, governmentality, and the triangle in four ethnographies.[1] Foucault differentiates between his analysis of government and the ethnographic study of the "witches' brew" of practices that actually occur in reality (Foucault 1991:81–2, cited by Li 2007:27). This chapter thus takes us from theory to practices as seen in ethnographies. My goal is twofold: to clarify Foucault's theoretical concepts in grounded examples, and to establish the usefulness of these three expressions of government for analyzing real conservation interventions. This chapter also underscores the importance of ethnographies, which alone can reveal the witches' brew of practices.

Each case study reveals multiple aspects of the triangle, though I selected the first two, by Roderick Neumann and Christophe Bonneuil, primarily to demonstrate sovereignty and discipline, respectively. They are extreme examples, but real ones that occurred widely, and still occur. The last two ethnographies, by Donald Moore and Tania Li, explicitly involve all aspects of the triangle. All four authors use Foucault to analyze their ethnographies. Moore and Li also analyze Foucault, his triangle, and the difference between his theory and grounded practices.

The triangle of government clearly exists in conservation. Displacing people from parks is sovereignty even when it's not violent. Excluding people from parks is also sovereignty. If people displaced from parks are resettled in buffer zones these resettlement schemes are usually disciplinary, involving regulations and some level of surveillance, if only to make sure people don't re-enter park territory. Participatory interventions designed to make buffer populations more amenable to their continuing exclusion are usually governmental: they may have to apply to participate, undertake trainings, or sign contracts. All this is done without the awareness that it constitutes government, or the understanding that sovereignty may undermine governmentality.

Three of these ethnographies are based on ethnographic fieldwork. Neumann carried out at least two periods of research in Tanzania: 1989–90 and 1997–2000. Moore carried out fieldwork in Zimbabwe in 1990–2. Li carried out fieldwork in Central Sulawesi, Indonesia, in the 1990s, spanning New Order and post-New Order periods. Bonneuil is an environmental historian.

Sovereignty

Roderick P. Neumann's study of violence in Tanzanian park conservation is a clear example of sovereignty. Neumann is a geographer and political ecologist.[2] He sees violent coercion in wildlife conservation in Tanzania, during the colonial period, after independence, and continuing in the 1990s alongside participatory approaches to conservation (Neumann 2001:313). He does not use Foucault's concept of sovereignty to analyze this violence, but we can productively do so. He uses Foucault's concept of discipline to analyze the participatory approaches of the 1990s, which I will examine in the next section.

Neumann's primary example of "state violence"[3] is displacement. In Tanzania, parks and protected areas were initially established and people displaced in the colonial period. Tanzania was a German colony until 1916, and a British one until 1961. The Germans set up the first game reserves and put the first hunting laws in place; the British maintained German laws and reserves and added to them, but were also constrained. The 1940 game bill, for example, says that nothing in the law "shall make it an offense for a native to hunt … for the purpose of supplying himself and his dependents with food," and any Africans born or residing within game reserves continued to own property (including pastures for livestock), and could "enter or reside within" reserves (Neumann 2001:309). Nevertheless, huge displacements occurred. The British moved 40,000 people out of the Selous Game Reserve, which had been established by the Germans, ostensibly for the control of sleeping sickness. Assuming that this massive displacement was coercive, it constitutes sovereignty, but its justification—the people's welfare—is governmental.

Independence in 1961 brought increased levels of violence, as the new nation needed tourist revenue in the parks and reserves, and then, in 1974, launched a massive "villagization" resettlement program to facilitate rural development (Neumann 2001:313, 322). Violence or the threat of violence continued to characterize national parks and protected areas. People caught hunting in parks could be beaten, raped, or shot as "bandits" by game officers organized like a paramilitary force; as recently as 1997 a shoot-on-sight directive was issued by a minister of natural resources and tourism (Neumann 2001:305, 320).[4] In 1989, in Operation Uhai, poachers were labeled as enemies of the nation, and 2,500 people were arrested (Neumann 2001:321).[5] The classifying of "poachers" as enemies of the nation, or even as "bandits," the paramilitary organization of game officers, the violence, and the shoot-on-sight directive all constitute sovereignty, with the park boundaries being treated like the territory of a medieval state. The nation of Tanzania coercively displaced its own citizens from state-owned territories, then used guards modeled on the military to punish transgressors it saw as enemies rather than citizens, sometimes taking their lives. I would add to Neumann's political-economic explanation for Tanzanian violence that a sovereign mentality is clearly operating. Its roots seem to lie in colonialism, but that history does not explain it.

Neumann (2001:323) writes that conservation professionals "would argue that … there are new conservation policies that involve the people in decision-making and that redistribute the benefits derived from wildlife tourism." Against this hypothetical argument, Neumann (2001:324) says that "state violence and the new community-oriented initiatives are integrated forms of social control designed to meet the needs and goals of international conservation organizations and the tourism industry." As we will see below, he sees the new initiatives as examples of Foucault's disciplinary power underwritten by the threat of violence. "The point," he writes, is that "coercion and violence … are not *replaced* by community participation but continue as ever-present threats that influence the behavior and decision-making of local community members" (Neumann 2001:327, author's italics). I think he could have gone farther, to suggest that disciplinary power in community projects may depend on a history of sovereign power inside parks.

Coercive displacement can also be seen in Bonneuil's, Moore's, and Li's ethnographies. I would argue that displacement for conservation constitutes Foucault's sovereignty because it excludes people from a territory and depends on coercion and often violence to do so.

Discipline

Christophe Bonneuil's[6] study of settlement schemes in tropical Africa during the late colonial[7] and early post-colonial period (1930–70) serves as an example of discipline. Bonneuil (2000:259) argues that science informed the "developmentalist" state. Prepackaged settlement schemes were imagined as laboratories where "development as experimental science" could be produced. The schemes were highly disciplinary:

> In these schemes, the land use, farming system, work, and life of the thousands of African tenants, brought from other regions, were strictly constrained by a disciplinary order.
>
> *(Bonneuil 2000:262)*

Fields and villages had to be "uniform" (they were in fact grids within grids) and schedules were "rigid" and centrally determined: "Farmers were working and living under permanent scrutiny and control" (Bonneuil 2000:267, 273n54). Settlement schemes had to be disciplined in part in order for them to serve as scientific laboratories:[8]

> Geometrization, simplification, standardization, and discipline ensured not only the social order and legibility sought by the state but also the experimental order necessary to produce expert knowledge.
>
> *(Bonneuil 2000:271)*

This allowed scientists to "domesticate" indigenous knowledge, to put it into terms that they could understand, circulate, store and sometimes rediscover. For example, scientific experimentation eventually resulted in demonstrating that traditional slash-and-burn cultivation worked, though their original form had been suppressed. Such schemes are now seen as "monsters" resulting in failure (Bonneuil 2000:280, and see Chapter 9). They are classic examples of the disciplinary power developed in European prisons, factories, schools, and armies and transferred to colonial rural development.

As we have seen above, Neumann uses Foucault's concept of disciplinary power to analyze the community-oriented buffer zone program put in place around the exclusionary and violent park space in Tanzania. The buffer zone, as he notes, has a history of the sort of villagization described by Bonneuil (Neumann 2001:323). In the buffer zone, violence is replaced by self-surveillance, as the responsibility for conservation shifts to villagers (Neumann 2001:325–6). Neumann (2001:326) argues (following Foucault's (1997) *Discipline and Punish*) that "the buffer zone plan serves as the 'discipline-mechanism' to create a different kind of peasant consciousness toward wildlife based on a schema of generalized surveillance." I would argue that these community projects are more governmentality than discipline. While surveillance typifies discipline, self-surveillance typifies governmentality. The intent to change peasant consciousness towards wildlife is also governmentality. These projects in many ways parallel projects in India that Arun Agrawal has argued are "environmental governmentality," as we will see in Chapter 10.

In any case, the sovereign power historically exercised inside parks may underwrite the power—discipline and/or governmentality—in surrounding community projects. Research on conservation organizations and their professionals is needed, however. How do they perceive past violence? Do they, knowingly or not, rely on it to attract participants for community programs?

Governmentality

Both Donald Moore[9] and Tania Li[10] apply Foucault's concept of governmentality to conservation and development on the ground. Both pair it with resistance, which is already a contrast to over-simplified governmentality literature that does not question its power. The focus of Li's *The Will to Improve* (2007) is on the agents of governmentality, those with the "will to improve"—who she calls "trustees"—and also on those with the agency to resist improvement in practice—"prickly subjects." This focus allows her to see governmentality as insecure in practice.

Moore's focus in *Suffering for Territory* (2005) is on those who resist, and have a history of resisting, multiple sources of power, in order to anchor their livelihoods in a particular place. Essentially people must wrest livelihoods out of landscapes that are both geophysical and "integrally entangled in power relations," including: a history as a colonial farm and as a site of anti-colonial resistance; the home of localized ancestral spirits; a colonial and national park and state forest; a past and current chiefdom, headman's territory, and rainmaking territory; a post-colonial resettlement scheme; historic wattle plantations and contemporary fruit plantations; and rain, which people recognize as having helped the anti-colonial resistance (Moore 2005:23, 25, 311, 317–19). Moore's understanding of power grounds theory about power in landscapes and their ecologies.

Li describes colonial Dutch and national Indonesian projects to improve the population, forests, and productivity of the land in Sulawesi, Indonesia. She sees the "will to improve," the title of her book, as characteristic of governmentality. The governmental rationale of improvement is constant in the colonial and national projects she reviews, but the techniques vary.

The Dutch held a liberal belief in retaining local culture and autonomy, but still used sovereignty-like displacement or military conquest as first steps in improvement. The Dutch missionaries actually recommended military conquest and forced resettlement after ten years of study; they saw "the sovereign's right to kill and command [as] necessary to set the conditions for improvement" (Li 2007:67–8). A Dutch missionary ethnographer and linguist carried out research in the Sulawesi hills to inform conversion, which they understood as necessarily retaining an "authentic cultural core" (Li 2007:67). They saw themselves as trustees, responsible for selecting cultural elements to be eliminated or retained (Li 2007:68). They planned to eliminate swidden (slash-and-burn) cultivation, feuding, headhunting, curing, burial and ancestral spirits, and local spirits (Li 2007:68). With these exclusions, "conversion would … emerge 'freely' from the people's own (altered) beliefs and aspirations" (Li 2007:68). Resettlement—a nicer word for displacement—would accomplish several of the planned eliminations. A colonial official similarly believed that military conquest would bring peace and enable economic development, "a process that would follow naturally so long as the Natives were granted sufficient autonomy" (Li 2007:71).

Li notes that these Dutch techniques mix governmentality and sovereignty. As I have said, the goal of improvement is governmental. The liberal belief in local culture and autonomy is also governmental; the subject of governmental power is a free subject. So is the ethnographic and linguistic research, which is only necessary because local culture and autonomy are deemed to be important.[11] But the Dutch interestingly used the results of this research to take a razor to local culture, separating those aspects to be excluded from those aspects to be retained. Because their goal included religious and agricultural improvement, local religion and agriculture were to be eliminated alongside local violence. Most of local culture ended up being treated in the way sovereignty treated lepers. Yet this stripping down was imagined to leave people ripe for governmental improvement.

The New Order national government was actually less governmental, carrying out no research, and holding no liberal belief in local culture or autonomy. They labeled uplanders as *orang asing* (foreigners), and treated them in a disciplinary way meant to overcome deficiencies (Li 2007:79). But Li argues that the rationale and strategy remained governmental: "They set new conditions from which, they calculated, improved conduct would follow" (Li 2007:94). New Order means were a hybrid of sovereignty and discipline for a governmental end.

Moore describes colonial British and national projects to improve the population, forests, and productivity of the land in Kaerezi, Zimbabwe. During the colonial past, Moore (2005:9) describes "a dangerous duet of illiberal and liberal rule," which combined violent conquest with discourses of imperial improvement. This sounds much like Li's case, but the British settled in Zimbabwe and established farms and ranches, which did not happen in Sulawesi. Kaerezians were violently expelled from land claimed by whites—Moore terms this "racialized rule"—and then transformed into tenants by "guiding their conduct and encouraging self-discipline" (Moore 2005:3). Whites managed, improved, and developed nature and natives (Moore 2005:13). This is sovereignty closely followed by governmentality, which Moore defines as "enlisting subjects in the project of their own rule" (Moore 2005:3).

Post-colonial Kaerezi displays all three modes of power. Nyanga national park and its annex are policed by armed patrols; park law criminalizes both hunting and harvesting as poaching. Villagers hunt deer and harvest firewood, poles, fruit, and reeds under the threat of violence. The state wants Kaerezians to move to "the lines" of a resettlement site, and occasionally sends in the army to torch out-of-place huts. The empty resettlement "lines"—they are actually marked in the landscape—represent extreme spatial discipline, separating fields and homes. A government cattle dip to benefit both citizens and the nation represents governmentality.

Subsistence production in this troubled landscape constitutes resistance to the state. Kaerezians open productive fields and build huts nearby in locations "wrong" in the eyes of the state. They choose to do so to cope with effects of structural adjustment: price inflation. Many had worked in the past, and they kept children in school while farming, including in college. Grown children, now professional, supplemented subsistence, sending money for school fees, seed, fertilizer, and clothes. Both the colonial and post-colonial states encouraged Kaerezians to enter the modern market. This they do, albeit on their own terms. In doing so, they are participating in the project of their own rule. This is governmentality. Subsistence production marks the limits of Kaerezians' willingness to rule themselves—the limits of governmental power.

There are also elements of governmentality in Neumann's and Bonneuil's projects. The buffer zone project in Tanzania required villages to volunteer (albeit in a context informed by coercion), and used contracts between villages and the government to shape the disciplinary spaces of their new lives. Self-surveillance was an odd mixture of discipline (everyone feels watched all the time) and governmentality (the intent was to create conserving subjects). Bonneuil's developmentalist colonial African states use disciplinary methods for a governmental purpose, to improve living conditions: "to turn African societies into objects of its cognitive apparatus and rationalizing interventions," informed by scientific experts, who reconceptualized environmental and health problems as "mere technical problems to be solved with appropriate expertise" (Bonneuil 2000:260, 265–6). Bonneuil (2000:281) ends his piece by saying that settlement schemes were laboratories where knowledge about African societies and environments and "a new governmentality" were "co-constructed."

The limits of power is one of Tania Li's primary underlying themes in *The Will to Improve*; it is also consistently present in her other published work. In Donald Moore's *Suffering for Territory* all types of power are insecure. Li (2007:10) states that governmentality "should be seen as a project, not a secure accomplishment." In particular, the critical separation between the agents of improvement, trustees, and its subjects is "a boundary that has to be maintained and that can be challenged" (Li 2007:7). Seeing power as a project rather than an accomplishment typifies both Li's and Moore's approach, and differs from authors whose use of Foucault is less ethnographic. The limits of power, in other words, are most apparent in practices. Neither Neumann nor Bonneuil provide enough ethnographic detail to make the limits to power visible in their cases.

Conclusion: Foucault's triangle

Governmentality in the contemporary world, certainly in conservation, exists alongside discipline and sovereignty. Conservation in Tanzania is a mixture of sovereignty-like coercion, spatial discipline, and governmentalized participation, and Neumann makes the point that the coercion and participation are actually integrated with each other. Bonneuil's developmentalist states use disciplinary methods for a governmental purpose: to improve living conditions. They also use coercion to move people into their disciplinary schemes. In Sulawesi the techniques used to practice the governmental goal of improvement (by Dutch and New Order governments) included the sovereignty of forced resettlement and exclusion of people from upland forests newly declared as national parks, and the disciplinary intensification of agriculture (Li 2007:61). In Zimbabwe people were resisting the discipline of resettlement, in spite of possible sovereign violence—but were otherwise improving themselves (which is governmentality), primarily through education.

Moore and Li both theorize the triangle. Moore (2005:7, 12) translates Foucault's triangle of sovereignty–discipline–governmentality into a "triad in motion," which is not stable, but shifting and contingent. Li (2007:13, 16) argues that practices of government "articulate elements of government, sovereignty, and discipline"; colonial means and objectives, for example, were "hybrid," "awkward amalgams," and all elements continue to be in play today. In Li the elements can be in contradiction. For example, colonial sovereignty (the right to rule based on a racial difference) was in contradiction with the will to improve (to make the colonized just like the colonizers) (Li 2007:14). In Li's terms Neumann's case is less an integration between sovereignty and governmentality than a contradiction, since participation in a violent context is not free participation. Li would also see Bonneuil's case as contradictory, denying the improving purpose of resettlement with the sovereign and disciplinary methods of resettlement and improvement.

In the implementation of conservation, the triangle of government clearly exists: most typically, peoples in the buffer zones of parks, having been displaced and excluded in a manner typical of sovereignty, are then exposed to interventions that may scrutinize them in a disciplinary way but also involve more governmental elements of participation like volunteering and contracts. Li argues that elements of sovereignty are in contradiction with elements of disciplinary and governmental power; I agree with this, and suggest that sovereignty seriously undermines more liberal approaches to improvement.

Political-economic power is not absent in any of these readings. Colonial states were extractive: during the colonial periods in Africa and Indonesia, people had to pay taxes and other fees, sometimes for the first time (see Li 2007:70). In the early national period, new states like Tanzania desperately needed sources of revenue and foreign exchange, like those from tourism in Neumann's case. Some officials are corrupt in Li. Li considers that excluding political-economic factors limits governmentality. Moore differs, arguing that capitalism must be seen within governmentality.

As I said in Chapter 5, I do not condone any type of government. I believe that conservation would be most effective without violence, surveillance, or the manipulations typical of "participation." Displacement and exclusion undermine everything else we try to do in conservation. Regulations, surveillance, and enforcement criminalize people who are usually just farmers. Acting on people's actions or conducting their conduct without knowledge of what they were doing before or what they need to do is both manipulative and disrespectful. What would conservation without government look like?

Notes

1. Three of the ethnographies are in Africa: Tanzania, tropical Africa, and Zimbabwe. The fourth is in Sulawesi, Indonesia, in Southeast Asia.
2. Neumann is a professor in the Department of International Relations at Florida International University.
3. Displacement can also be called resettlement, especially in cases where displaced households are moved into resettlement schemes. Sovereignty is only relevant to the displacement part of the process; resettlement schemes are often disciplinary, but may also be governmental.

4. The levels of violence in Neumann's case may seem extreme. But paramilitary training and arming of personnel who monitor protected areas and informal shoot-to-kill policies occur in other countries (Moreto 2019:2).
5. Neumann's analysis of this violence is primarily political-economic. "Historical peaks in violence," he writes, "often are associated with a reaction by the state to some threat to its control over revenue generated from wildlife, its domestic political legitimacy, its moral standing in the international sphere, or some combination of these" (Neumann 2001:322). By moral standing, he means bad international publicity about conservation, blaming international conservation organizations (Neumann 2001:323). In contrast, he links poaching to needed access to livelihood resources, while mentioning that organized transnational networks of illegal trade in elephant ivory and rhino horn emerged in the 1980s (Neumann 2001:309, 320). The winners are thus international conservation organizations and the nation. The losers are local people. I find this unsatisfactory as an explanation. I want to know more about the transnational illegal ivory and rhino horn trade. Surely that trade carries a great deal of the blame. How did it intersect with local people? I also want to know more about conservation organizations in Tanzania. Did they actually have the ear of policy-makers in Tanzania? Did they incite or even turn a blind eye to violence? Ethnographies of the illegal trade and of conservation professionals are needed.
6. Bonneuil is a senior researcher in the history of science, science studies, and environmental history at the Centre Alexandre Koyré (Centre National de la Recherche Scientifique) Paris. His research explores the co-evolution of ways of knowing and ways of governing nature and the earth.
7. Disciplinary settlement in Africa begins later than disciplinary government in Europe. This raises the question of whether the colonial nations of Europe applied different thinking about government to their own citizens and to the peoples in their colonies. Were colonies, like prisons, schools, armies, and factories, disciplinary spaces long after the historical transition to governmentality?
8. The connection that Bonneuil draws between discipline and science is deeply interesting. It suggests that wherever scientists are using local people to help them carry out research they may be relying on some disciplinary power in order to do the science right. See also Chapter 7 on Latour's concept of metrologies.
9. Donald S. Moore is associate professor of anthropology at Berkeley.
10. Tania Murray Li is professor of anthropology and Canada Research Chair in the Political-Economy and Culture of Asia at the University of Toronto. She has two books that are essential reading for environmental anthropologists: *The Will to Improve* (2007) and *Land's End* (2014).
11. Another official (in 1935) applied a cost–benefit equation to arrive at a policy (Li 2007:76); this is thoroughly governmental.

References

Bonneuil, Christophe 2000 Development as Experiment: Science and State Building in Late Colonial and Postcolonial Africa, 1930–1970. *Osiris* 15:258–81.

Foucault, Michel 1977[1975] *Discipline and Punish: The Birth of the Prison*. Alan Sheridan, trans. New York: Vintage Books.

Foucault, Michel 1991 Question of Method. In *The Foucault Effect: Studies in Governmentality*. Graham Burchell, Colin Gordon and Peter Miller, eds. Pp. 73–104. Chicago: University of Chicago Press.

Li, Tania Murray 2007 *The Will to Improve: Governmentality, Development, and the Practice of Politics*. Durham, NC: Duke University Press.

Li, Tania Murray 2014 *Land's End: Capitalist Relations on an Indigenous Frontier.* Durham, NC: Duke University Press.

Moore, Donald S. 2005 *Suffering for Territory: Race, Place, and Power in Zimbabwe.* Durham, NC: Duke University Press.

Moreto, William July 23, 2019. "Nobody Studies Law Enforcement (Rangers)"—But We Need To. *Oryx: The International Journal of Conservation.* https://oryxthejournal.org/blog/?s=Moreto. Accessed November 11, 2019.

Neumann, Roderick P. 2001 Disciplining Peasants in Tanzania: From State Violence to Self-Surveillance in Wildlife Conservation. In *Violent Environments.* Nancy L. Peluso and Michael Watts, eds. Pp. 305–27. Ithaca, NY: Cornell University Press.

7

STATES AND CENTERS, SIMPLIFYING AND CALCULATING

Introduction

Thus far, this book has examined approaches to conservation based on the seminal ideas of Foucault about power and government. Other ideas, also more or less outside of political ecology,[1] have been important in the anthropology of conservation. These authors read well with Foucault, as valuable supplements to ideas we have already examined in this book. In this chapter I examine the ideas of James C. Scott, Bruno Latour, and Anna L. Tsing. All three write about the environment. They provide more tools for our tool-box.

In *Seeing Like a State*, James Scott (1976) argues that states simplify or render legible in order to govern, and that modern states base their simplifying interventions on a valorization of science and technology. In other words, there is a significant relation between legibility, scientific and technical knowledge, and the government of society and the environment. Bruno Latour talks about science, not states. He is also often cited alongside Foucault (e.g., Braun 2000; Mitchell 2008), and adds the actual technologies and practices of rendering legible and calculating, and how they extend power across space, to Scott. Anna Tsing explicitly expands the sorts of knowledge that inform simplifying interventions to include the "neutral" knowledge through which the environment is "explored, managed, converted, and conserved," which depends on calculative practices of efficiency, natural capacity, and profit margins. She brings conservation into theories of simplifying. She also gives us a lyrical ethnographic glimpse of the social and cultural matters that such knowledge erases.

I argue that Latour, Scott, and Tsing incorporate an ethnographic understanding into theory about power. In Scott, practical knowledge makes state schemes work, but is discursively erased; local society, practical knowledge, and nature, as he knows from ethnographic research in Malaysia, resist complete simplification (see Chapter 14). In Latour, the practices of collecting inscriptions and analyzing them should be emphasized; building networks of allies and extending the laboratory into the outside world, based on his research in laboratories, is how power grows. Tsing argues that the relationship between the Meratus people, trees, and bees, which she knows from living with them, is an example of a set of Meratus ideas and practices that offer a "challenging alternative to European models."[2]

Scott

The thesis of James C. Scott's[3] 1998 book, *Seeing Like a State: How Certain Schemes to Improve the Human Condition Have Failed*, is that the high-modernist ideology and simplifying abstractions (rendering legible) of states have the power to reshape society and nature. States also attempt to exclude the local, practical knowledge on which they depend. Fortunately, local society, practical knowledge, and nature resist complete simplification. He clarifies:

> I am emphatically not making a blanket case against either bureaucratic planning or high-modernist ideology. I am, however, making a case against an imperial or hegemonic planning mentality that excludes the necessary role of local knowledge and know-how.
>
> *(Scott 1998:6)*

Scott contrasts the hedgehog, who knows one big thing and is thus systematic and synoptic but also static and myopic, with the fox, who knows many things. Foresters and cadastral officials are hedgehogs, as is the state itself; naturalists and farmers are foxes (Scott 1998:45–6).

Power in this thesis is the power in *seeing* like a state, in legibility and simplification—even when that state is not authoritarian and coercive. Seeing like a state is seeing like a hedgehog. State simplifications are interested, utilitarian facts; written documentary facts; static facts; aggregate facts; and standardized facts (Scott 1998:80). State functionaries, in fact the functionaries of any large organization, "see" through the "simplified approximations of documents and statistics: tax proceeds, lists of taxpayers, land records, average incomes, unemployment numbers, mortality rates, trade and productivity figures, the total number of cases of cholera in a certain district" (Scott 1998:76–7).

The market works much like the state in rendering legible and reducing everything to numbers, Scott (1998:8) writes: "Large-scale capitalism is just as much an agency of homogenization, uniformity, grids, and heroic simplification as the state is." "Commercial logic and bureaucratic logic were … synonymous" (Scott 1998:18). But for capitalists, "simplification must pay" (Scott 1998:8). Surnames, for example, serve both state needs for administration and taxation and capitalism's needs, and of one Scott's examples, Paris, is simplified for health, economic efficiency, and military security (Scott 1998:62, 67–8).

Scott's simplifying state often sounds like Foucault's disciplinary state: "The utopian, immanent, and continually frustrated goal of the modern state is to reduce the chaotic, disorderly, constantly changing social reality beneath it to something more closely resembling the administrative grid of its observations" (Scott 1998:82).[4] This similarity probably stems from the fact that both Foucault and Scott wrote about the early modern state, when administrative ordering was new. They both suggest that the pure dream, whether for order or simplicity, does not work. In Foucault (2007), governmentality follows the realization by the physiocrats that minute state regulation cannot prevent famine; in Scott, schemes to improve the human condition fail without practical knowledge (see Chapter 13). *Normalbaüme* is one such failure.

The *normalbaüme* was the perfected forest of German scientific forestry, excluding everything natural except trees. This suggestively parallels conservation when it excludes people and their livelihood activities, especially when the park's ecology developed alongside those activities. The fiscal, revenue-yielding, scientific forest—*normalbaüme*—replaced "the actual tree with its vast number of possible uses" with an abstract tree (Scott 1998:12). The state's "tunnel vision" missed the importance of both the social and the ecological: "nearly everything touching on human interaction with the forest" and "the forest as habitat" (Scott 1998:13). Nature was reduced to a single natural resource. This led eventually, after about a century, to a devastating loss of resilience in *normalbaüme*: "A whole world lying 'outside the brackets' [e.g., the exceptionally complex processes of soil building] returned to haunt this technical vision" (Scott 1998:20). However, by that time German forest science was "hegemonic" in the world, and German scientists pioneered efforts to remedy *normalbaüme* problems, adding back some of the things they had excluded (Scott 1998:19, 21).

Scott compares the scientific forest to rationalized measures, the "metrical revolution," and the cadastral survey and mapping, which try to render legible the illegible local practices of measurement and landholding. The "metrical revolution" in France established nation-wide standardized measures as part of the French Revolution, because it eliminated the feudal power held by aristocracy and clergy, which had established local measures. It also allowed a new sort of state building, because "no effective central monitoring or controlled comparisons were possible without standard, fixed units of measurement" (Scott 1998:30). It was, in fact, part of a larger project to create a uniform, homogeneous, and rational citizen, and thus "a single national society perfectly legible from the center" (Scott 1998:32). Feudal power gave way to a newly centralized power that depended on the legibility of local measures and landholding from the center. The metrical revolution makes clear a critical aspect of the power of simplification: it makes things legible to a center. Simplification constitutes a positional advantage; it establishes a privileged vantage point (Scott 1998:78–9).

The cadastral map and land titling also simplified and rendered legible to serve a center of power: "Just as the flora of the forest were reduced to Normalbaüme, so the complex tenure arrangements of customary practice are reduced to freehold, transferrable title" (Scott 1998:36). The "fiscal feudalism" of collective taxation was in this way eliminated, and the fiscal take of the state, based on individual landholders, was rationalized. Cadastral maps are simplifications:

> The completeness of the cadastral map depends, in a curious way, on its abstract sketchiness, its lack of detail—its thinness.... What lies inside the parcel [of mapped land] is left blank—unspecified—since it is not germane to the map plotting itself.
>
> *(Scott 1998:44)*

What lives inside the blank is the fox and what it knows, local knowledge, local society, and local ecology.[5]

Such mapping and land titling projects were often used to rationalize land holdings, such that each household would own a single piece of arable land rather than many small strips of land spread around the village territory. The rationalized maps are reminiscent of the resettlement schemes described by Bonneuil in Chapter 6; Scott devotes a whole chapter of *Seeing Like a State* to villagization in Tanzania. This adds a second aspect to simplification (alongside making things visible to a center): "The categories used by state agents are not merely means to make their environments legible; they are an *authoritative* tune to which most of the population must dance" (Scott 1998:83, author's italics). Nature and peasants themselves, not just their names and measures and maps, were simplified. Fortunately, local society could in some cases "modify, subvert, block, and even overturn the categories imposed upon it" (Scott 1998:49).

States rationalized land holdings for a reason; for example, to transform peasants into entrepreneurs who would embrace scientific agriculture (another simplification; see Scott's Chapter 8). Scott terms these aspirations "authoritarian high modernism." They combine the authority of science with the goal of improving the human condition in a way that closely parallels the *normalbaüme* (Scott 1998:93). This goal is reminiscent of the goal of governmentality in Foucault, to improve the welfare of the population. It adds a third aspect to the power of simplification (alongside visibility to a center, and reshaping nature and society): a belief that a state can and should improve both nature and society.

In sum, Scott's concept of power is the power of "seeing" like a state, the power in simplification itself and in the way simplification creates legibility. Simplified facts are legible to a center. That center, the state, can turn these simplified facts around and reshape nature and society. And early modern states believed that they could and should improve nature and society, using science. Scott's concept of power is close to Foucault's concept of disciplinary power, but shows some aspects of governmentality.

Scott's focus on a simplifying gaze and the legibility it demands and creates resonates with Foucault's discursive power as well. But Scott's grounding in peasant studies and his field research with peasants mean he is always aware of what the simplifying gaze does not see,[6] and he understands the power of ecologies and knowledges that are illegible to the state's gaze.

Scott's insights about power are applicable to conservation. Conservation organizations exist to improve nature and society using science. Like high-modernist states, they displace people and their traditional uses of nature, and attempt to replace them with something more modern in both the scientific and economic senses. They do not appreciate how local society limits their ability to do this.

Latour

Bruno Latour[7] writes about science, not states. But science includes the administrative sciences that build Scott's states: "a stable state of society is produced by the multifarious administrative sciences exactly like a stable interpretation of black holes is provided by astronomy" (Latour 1987:256). Disciplinary power also served science in Bonneuil's case study of tropical African resettlement schemes (see Chapter 6).

Latour explains the birth of modern science in the sixteenth century and its significant effects primarily by the development of writing and imaging craftsmanship, "inscriptions" that are mobile, immutable, flat, can be scaled down, are reproducible, re-combinable, superimposable, insert-able into written text, and can be translated back into three dimensions with geometry: that is, "immutable mobiles." The other factor is agonistic encounters, which are contests to win allies and make statements true. The purpose of inscriptions and cascades of inscriptions is "to swing the balance of power" in agonistic (combative) encounters, thus turning "an incredible statement into a credible one" (Latour 1986:4) This is "the way we argue, prove and believe"; "the way in which someone convinces someone else to take up a statement, to pass it along, to make it more of a fact, and to recognize the first author's ownership and originality" (Latour 1986:5, 17). In Latour (1987:232) scientists and engineers "travel inside narrow and fragile networks," making traces circulate better by increasing their mobility, speed, reliability, and combinability. The results of "building, extending and keeping up these networks is to act at a distance," sometimes dominating the periphery (Latour 1987:232). The combination of inscriptions and network will become actor–network theory,[8] which we will link to assemblages in Chapter 14.

Latour's (1986:12, 14) question in "Visualisation and Cognition" is how to dominate on a large scale. The answer concerns a second level of agonistic encounter, in which a "center" emerges. Centers dominate by "paper shuffling" and concentrating files: "It is hard to overestimate the power that is gained by concentrating files"; "a man whose eye dominates records … may be said to dominate"; "the few may dominate the many" (Latour 1986:26). This is similar to Scott's concept of the simplifying state, which also depends on written, static, aggregate, standardized documents and statistics. In Latour (1987:218), the first step in this domination is cycles of accumulation: "The accumulation [of inscriptions] that will generate an asymmetry [of power] hinges upon the possibility for some traces of travel to go back to the place that sent the expedition," which thus becomes a center. Translation into new forms—numbers, money, maps—allowed this accumulation in centers. The cycle of accumulation allows a point to become a center by acting at a distance or dominating at a distance many other points (Latour 1987:222). Scott emphasized the act or the process of simplifying; Latour emphasizes the things moving to a center (the mobiles), their movement, and their accumulation.

In Latour, the process continues once immutable mobiles reach the center, and that process is simplification, seen a bit differently than Scott. New calculations were developed in centers to analyze—essentially to simplify—the accumulation of inscriptions: Latour (1987:233) calls it "mopping up" the inscriptions. This is statistics, which he calls a "cascade" of inscriptions: "This cascade of fourth, fifth and nth order inscriptions will never stop, especially if the population, the computers, the profession of demography, statistics and economics, and the Census Bureau all grow together" (Latour 1987:234). In the age of "big data," we know it has not stopped. Equations enhance combinability and transform centers of accumulation into centers of calculation, centers inside centers (Latour 1987:235, 239).[9]

In Latour's work the cycle of accumulation is centripetal, but science is also applied centrifugally. This occurs through the "progressive extension of a network … *outside* of the networks that gave birth to them" (Latour 1987:248–9, author's italics). This is what Latour calls "metrology": "the name of this gigantic enterprise [technoscience] to make of the outside a world inside which facts and machines can survive" (Latour 1987:247–57). Metrology makes "action at a distance possible" (Latour 1987:254). The word "metrology" literally means the science of measurement. This is the French metrical revolution in Scott. Scott was primarily interested in what happened when state simplifications were imposed on people. But unlike Latour this takes Scott to the necessary role of local forms of knowledge and the failure of simplifications that succeed in excluding it.

Like Scott and Foucault, Latour moves easily to power in the economy: "There is not a history of engineers, then a history of capitalists, then one of scientists … then one of economists. Rather, there is a single history of these centers of calculation" (Latour 1986:29). Capitalism itself, he suggests, may be another process of immutable mobilization (Latour 1986:28).

Latour's focus on the accumulation and simplification of immutable mobiles and metrology adds a layer to the understanding of how scientists and bureaucrats build influence and work to extend it into the world. But his understandings of power itself, of agonistic encounters and centers of accumulation, adds less to Foucault and Scott. Foucault thought a great deal more about power relations, I think, and was largely uninterested in centers. Scott thought a great deal more about centers, and was largely uninterested in power relations. And Scott is the only one of the three to think about peripheries, about the blank spaces in maps, and about the potential of people to resist disciplinary power and metrology.

Tsing

Anna Tsing[10] gives us a deep view of the periphery based on ethnographic research. What she shows is there are culture and society, through which people know nature. It is the opposite of states and centers, simplifying and calculating. The complexity Tsing captures on the periphery is illegible from centers, and it is certainly irreducible to numbers. Culture and society see resource use as a matter of "aesthetics in its broadest sense: feelings, sensibility, moral vision": meanings (Tsing 2003:24). "Natural resources are cultural resources" (Tsing 2003:28). And the forest is "always a *social* landscape, not a 'wild' place" (Tsing 2003:37, author's italics).

Tsing contrasts knowing nature through culture and society to knowing it through economics and ecology. Economists and ecologists know nature as "neutral matters of efficiency, natural capacity, and profit margins," which allow it to be "explored, managed, converted, and conserved" (Tsing 2003:24). In other words, these kinds of knowledge simplify, quantify, and govern the environment. They have power, too, a power that resembles Foucault's power/knowledge or Scott's simplifying—the power to "hide." In Tsing's (2003:24) words:

> This convention of thinking hides, but does not stop, the making and affirming of cultural frameworks in all human activities. It also hides attempts to export particular cultural frameworks globally under the guise of neutral knowledge.

People, local and economists and ecologists alike, "know nature through cultural frameworks" (Tsing 2003:35). Even more seriously, international economic and ecological projects have a "blinding magic" that separates scientists, activists, and policy-makers from local insights about "the long-term coexistence of people and forests" (Tsing 2003:26). The problem lies not in the models of economy and ecology themselves, but in their power to hide complexities and alternatives—to appear, in Foucault's terms, to be the "truth." The local in Tsing are the Meratus people of Kalimantan, Indonesia.

Tsing uses bees and their honey to contrast simplifying power to the complex ways that Meratus culture and society interact with nature, to give a glimpse of what "blinding magic" is blind to. We can use a passage from Scott's (1998:3) "Introduction" in *Seeing Like a State* to represent the way that simplifying power has shaped bees and honey:

> A homely analogy from beekeeping may be helpful here. In premodern times gathering honey was a difficult affair. Even if bees were housed in straw hives, harvesting the honey usually meant driving off the bees and often destroying the colony. The arrangement of brood chambers and honey cells followed complex patterns that varied from hive to hive—patterns that did not allow for neat extraction. The modern beehive, in

> contrast, is designed to solve the beekeeper's problem. With a device called a "queen excluder," it separates the brood chambers below from the honey supplies above, preventing the queen from laying eggs above a certain level. Furthermore, the wax cells are arranged neatly in vertical frames, nine or ten to a box, which enable the easy extraction of honey, wax, and propolis. Extraction is made possible by observing "bee space"—the precise distance between the frames that the bees will leave open as passages.... From the beekeeper's point of view, the modern hive is an orderly, "legible" hive allowing the beekeeper to inspect the condition of the colony ... judge its honey production ... and, above all, extract just enough honey ... to ensure that the colony will overwinter successfully.

Tsing (2003) tells us that the Meratus don't keep bees; they hunt honey. Meratus trade with bees rather than domesticating them. A particular Meratus male grooms a named "honey tree" by cleaning off vines and clearing growth around it. He has rights to this tree that he expects even timber companies to respect, though they don't. Honey-tree rights are charismatic, based on narrated management and an imagined relationship with particular bee swarms, whose migratory agency in the forest, the Meratus believe, resembles their own. On the moonless night the honey is hunted, a young man climbs the 100-foot-or-more tree using bamboo stakes, with a sparking palm hair torch, while singing songs to the bees; the bees follow the sparks and the climber harvests the honey. He gets to be the organizing center of a social circle and to have a moment of fame; honey abundance, the Meratus believe, is attracted to Meratus prowess. The product of this charisma and prowess, the honey, must be shared equally and widely. Tree rights are individual; honey rights common. This allows common needs to be met at the same time that individuals "deploy their most entrepreneurial talents as managers of social and natural resources" (Tsing 2003:48–9). Meratus honey hunting is exponentially more difficult than even straw hives. But this difficulty allows it to be filled with social and cultural meaning. It cannot begin to be understood without an understanding of ideas of individual ownership versus common rights, the basis of leadership and its responsibilities, the imagination of nature and its relation to people, etc. Extracting honey from a modern, legible hive doesn't require meaning. More accurately, meaning probably does accompany modern bee keeping, but it is seen as outside the business of honey production, which is a neutral matter of efficiency, natural capacity, and profit margins.

Tsing (2003:49) frames her analysis of Meratus honey hunting as crafting tools for "respectful multi-cultural conversations" on tropical forest management: "I am suggesting that attention to local cultural ecologies may give us more tools to think about forest conservation in regions of human settlement." "The relationship maintained between people, honey trees, and bees," Tsing (2003:48) explains, "is one promising example of the kind of open-ended, semi-autonomous management relations that make it possible for people and forests to live together over time" and "offers a challenging alternative to European models." She argues that we must improve national and international frameworks for listening (Tsing 2003:53n17).

Power in Tsing is a magic that blinds ecologists and economists (and conservationists) to the fact that their own approach to nature is also cultural, and it blinds them to local natural history: a nature filled with the cultural and the social. They are blind to the complexities and aesthetics of local human–nature relations, on the peripheries of centers and states. Scott and Latour, and Foucault, analyze that blinding magic; Tsing's writing gives us its opposite, with open hands, over and over again.

Conclusion

This chapter has added several tools to our tool-box. First, Scott's concept of power is the power of "seeing" like a state, the power in the way simplification creates legibility to a center that can be made to reshape nature and society and is accompanied by the belief that nature and society should be improved in this way. Scott thinks about peripheries, about the blank spaces in maps, and about the potential of people to resist things like Foucault's disciplinary power and Latour's metrology. Like states, conservation organizations improve nature and society using science, sometimes displacing people and their traditional uses of nature and replacing them with scientific and economic improvements. Second, Latour's focus on the accumulation and simplification of immutable mobiles and metrology adds understandings of how scientists and bureaucrats control peripheries and extend that control out into the world. This is also applicable to conservation organizations that collect data in Western centers and use it to improve conservation in peripheries. Finally, Tsing's power is a magic that blinds ecologists and economists, and thus conservationists, to the fact that their approach to nature, too, is cultural, and also to local natural history: in which nature is cultural and social. This suggests that conservationists are blind to the complexities and aesthetics of human–nature relations in the peripheries.

All three authors link power to the economy. For Scott, the market works much like the state in rendering legible and reducing everything to numbers. For Latour, scientists and capitalists share the same history in centers of calculation and metrology. Tsing condemns economists along with ecologists for being blind to the cultural character of knowledge, especially knowledge of the forest.

I have argued that all three of these authors are ethnographers. This is very evident in Scott, whose perspective, even when writing about states, is from the periphery. But it is most evident in Tsing, because the complexities that she describes and valorizes as important to conservation can only be seen in ethnographic fieldwork, and can only be read in ethnographies.

Simplification itself is not evil—we all have to simplify all the time. But one cannot simplify without valorizing the simplifications and their metrics, and excluding—in fact blinding oneself to—complexities that may matter. Simplifications lend themselves to claims of truth. Complexities don't. If a simplification comes to be seen as truth, its "blinding magic," its discursive power, increases. Complexities do not disappear: they live on in the blank spaces on maps, the bugs and fungi in forests, the things that cannot be counted, the margins of states, and in ethnographies. What is problematic about simplification is its potential to govern our thinking such that complexities are erased from our minds, because complexities will emerge, in practice, in the implementation of our simplified policies. And if Scott and Tsing are right, we need them—and should be learning from them.

Notes

1. Scott's *Moral Economy* and *Seeing Like a State* are often used by political ecologists (e.g., Robbins 2012). Latour entered the political ecology lexicon with postmodernism, along with Foucault, though Robbins (2012:77) sees him as an awkward fit. To my knowledge, Tsing's writing is not used by political ecologists.
2. Scott and Latour, as far as I could find, do not cite each other; Tsing cites Latour and knows but does not cite Scott in this 2003 book.
3. James C. Scott is an American political scientist and anthropologist by adoption, probably because he has championed qualitative research over quantitative. He has been teaching at Yale since 1976 as the Sterling Professor at Yale and director of the Agrarian Studies program; before that he taught at the University of Wisconsin–Madison. Scott is sometimes said to be a Marxist, but is more accurately a fan of Karl Polanyi.
4. There is actually a note referencing Foucault's governmentality in Chapter 1 in Scott (1998:14n7).
5. In conservation interventions, private land titling threatens an array of social arrangements that may give neighbors and kin rights to things like tree fruits on land they do not "own."
6. This is like Ferguson's 1994 anti-politics machine (the development machine makes politics vanish, described in Chapter 3), which was based on Foucault, and also made projects fail.
7. Bruno Latour is a French philosopher and sociologist–anthropologist known for science and technology studies and actor–network theory.
8. It is related to Foucault's discursive power, though Foucault stays at a more general level (e.g., how truth is attached to a statement). In Foucault, power lies in the statement (its power as "truth"), not in the power of any author. When statements are taken up and passed along (which I like), do they carry with them recognition of the author's ownership and originality? This is a difference between the two. In Foucault, statements are authorless.
9. Foucault links things like statistics to governmentality, but he is focusing on the new mentality (optimizing, balancing costs and benefits). Latour's combinable immutable mobiles historically span Foucault's discipline and governmentality periods. We might compare Latour's distinction between (1) accumulated mobiles and (2) their 4th to nth degree inscriptions (cascades of inscriptions and calculations) to Foucault's distinction between (1) the disciplinary disposition of people and things into tables and grids and (2) the securing, through the science of statistics, of the economy and the population.

10. Anna Lowenhaupt Tsing is a professor in the anthropology department at the University of California at Santa Cruz, and author of *Friction* (2005) and *The Mushroom at the End of the World* (2015).

References

Braun, Bruce 2000 Producing Vertical Territory: Geology and Governmentality in Late Victorian Canada. *Ecumene* 7(1):7–46.

Ferguson, James 1994[1990] *The Anti-Politics Machine: "Development," Depoliticization, and Bureaucratic Power in Lesotho.* Minneapolis: University of Minnesota Press.

Latour, Bruno 1986 Visualisation and Cognition: Drawing Things Together. *Avant: Trends in Interdisciplinary Studies* 3(T):207–60.

Latour, Bruno 1987 *Science in Action: How to Follow Scientists and Engineers through Society.* Cambridge, MA: Harvard University Press.

Mitchell, Timothy 2008 Rethinking Economy. *Geoforum* 39:1116–21.

Robbins, Paul 2012 *Political Ecology: A Critical Introduction*. Second Edition. London: Wiley-Blackwell.

Scott, James C. 1976 *The Moral Economy of the Peasant: Rebellion and Subsistence in Southeast Asia.* New Haven: Yale University Press.

Scott, James C. 1998 *Seeing Like a State: How Certain Schemes to Improve the Human Condition Have Failed.* New Haven: Yale University Press.

Tsing, Anna Lowenhaupt 2003 Cultivating the Wild: Honey-Hunting and Forest Management in Southeast Kalimantan. In *Culture and the Question of Rights.* Charles Zerner, ed. Pp. 24–55. Durham, NC: Duke University Press.

Tsing, Anna Lowenhaupt 2005 *Friction: An Ethnography of Global Connection.* Princeton: Princeton University Press.

Tsing, Anna Lowenhaupt 2015 *The Mushroom at the End of the World: On the Possibility of Life in Capitalist Ruins.* Princeton: Princeton University Press.

8

ARTICULATIONS BETWEEN KNOWLEDGES IN ETHNOGRAPHIES

Introduction

My purpose in this chapter is to make us think about articulations[1] or connections between Western (global) and indigenous (local)[2] knowledges or discourses as relations of power. This chapter thus returns to discourses or knowledges to consider articulations that reveal the play of power *between* knowledges. In all three cases, Western or state power ends up dominating the articulation in some way, but the three cases differ in how this occurs.

This chapter will examine a history and two ethnographies: a connection between Western and local knowledges before the two were opposed, an articulation in the early 1990s when "indigenous knowledge" had emerged as a self-conscious discourse, and the purposive incorporation of indigenous knowledge into environmental management. The historical example by Richard Grove will show, against what many believe, that there was a time when local doctors on the Malabar coast of India knew a lot of things that their Dutch colonial governors did not. The story of how this history disappeared is the story of power in this case, which I will use Foucault and Latour together to analyze. Grove argues that Western botany, medicine, trade, and even conservation owe a lot to indigenous knowledge. The emergence of the idea of "indigenous knowledge" in the 1990s in Indonesia, as told by Tania Li, demonstrates the discursive power of "indigeneity," and its limits. In her ethnography, local people and non-governmental organizations (NGOs) in Sulawesi, Indonesia, deliberately created "a reified body of 'indigenous environmental knowledge' from the pool of everyday local knowledge and practice," in order to claim land rights. Such claims can succeed, but they force people to articulate their identity in terms recognized by government agencies, which will constrain how they can exercise their rights.

Finally, Paul Nadasdy uses Latour to think about how the integration of indigenous knowledge into cooperative natural resource management can have effects of power. In his ethnography, a project to integrate scientific and traditional environmental knowledge for environmental co-management in the Yukon fails. Using Latour, Nadasdy argues that a mandated integration of traditional ecological knowledge (TEK) and science is actually reinforcing science, and reinforcing the actual power that science has over First Nations' natural resources in the real world.

Before the opposition between Western science and indigenous knowledge

Most of us have forgotten that Western botany, medicine, and even conservation owe a lot to indigenous knowledge—and that very forgetting marks the power working on this historical articulation. The forgotten history that Grove documents should also establish the fact that Western scientific knowledge and local or indigenous knowledges are not actually separate things.

In Richard Grove's[3] case, a past nearly forgotten by history demonstrates that Western botany, medicine, trade, and even conservation owe a lot to indigenous (local, and lower caste at that) knowledge; in other words, indigenous sources contributed to (1) Western scientific and medical knowledge, and to (2) the Western domination of global trade and thus colonialism.

A historical connection did occur: early in the colonial period, in India, knowledge from a low-caste physician (Achuden) and his inherited books and papers were incorporated into a Dutch publication, the *Hortus Malibaricus*, authored by the Dutch physician (and governor of the Dutch colony of the Malabar coast of India) van Rheede. This connection was highly significant. The Dutch publication contributed to the Swedish botanist Linnaeus, who developed the modern system of taxonomy Grove 1996:139), and (together with gardens and publications of Dutch colonial centers in the Cape of Africa and Indonesia) "provided both the intellectual and informational base and the model for subsequent botanical developments in the British and French colonial empires of the [eighteenth] century" (Grove 1996:140). Thanks to Grove (and to van Reede himself, who recognized Achuden's contribution) we know about this connection and can trace its influence. But it is probably safe to say that few contemporary physicians or botanists—or conservationists—are aware of it; essentially indigenous knowledge was absorbed into sciences now identified as Western. A connection, over time, became an absorption.

This absorption of local knowledge is an indicator of power. Achuden and van Reede were low-caste local and colonial governor, respectively. But they were also both physicians, and in that role Achuden had knowledge that van Reede lacked and was looking for. Van Reede recognized Achuden's role in the *Hortus*, a fact that suggests that he respected him. It is the forgetting of this original moment of connection that is most interesting in terms of power. Western science and medicine in time came to be seen as inherently "Western," and as "truth" vis-à-vis local knowledges. This is the process that Foucault has described, of a knowledge accumulating the power of truth, establishing itself as superior to all other knowledges, as an objective knowledge of reality, and as global—when its historical origins are local, cultural, and political. This forgetting of the cultural character of science is also what Tsing criticizes in Chapter 7.

Grove adds that the original articulation affected not only knowledge, but economic and political domination in the real world. Economic botany is important to this history, too. Van Reede's publication contributed to the monopolies of the Dutch East India trading company (the VOC), which eventually became the Dutch colonies in Southwest India and Indonesia. This pattern was also followed by the British and the French trading companies. A connection became an economic and political absorption also, as Dutch traders collected local plants in the colonies for sale in Europe, eventually transforming the homes of these plants into colonies.

How did this happen? A hortus is a dry garden, a herbarium. Van Reede's *Hortus Malabaricus* translated dried plant specimens into drawings, supplemented them with text about uses, and published them in Leiden. A hortus is an immutable mobile as theorized by Latour (see Chapter 8): the publishing of this and other, similar books meant that originally indigenous knowledge was put into the hands of individual physician-botanists in Europe. The Dutch, and then the English and French, created botanical networks of books, colonial botanical gardens, and physician-botanists located in colonies. This informed colonial economic extraction of natural resources that were otherwise unknown. This is Latour's cycle of accumulation. Connections between indigenous and European physicians, fixed and reproduced as publications,[4] transformed tropical botany into both Western knowledge and eventually Western economic and then political power. Grove's analysis suggests that knowledge is primary, not economic and/or political power. In Latour's terms this extraction was only possible because of the translation of plants through paper, drawing conventions, and printing. The extraction of knowledge, dependent on the development of immutable mobiles, necessarily preceded economic extraction.[5]

Some of Achuden's knowledge also existed in the form of a collection of books and papers, but this collection was not an immutable mobile. Achuden was a traditional doctor and Ayurvedic medical practitioner, an occupation passed from father to son—"along with bulky collections of books and papers, containing hundreds of years of accumulated, medico-botanical knowledge" (Grove 1996:138). This collection was searched for in the 1980s, but permanently lost (Grove 1996:139). Achuden's passed-down collection lives on only in van Reede's *Hortus.*

Grove's ending point is not colonialism, however, but conservation. Knowledge of plants, together with their extraction, made it possible to think about the impact of extraction on the environment, a consciousness that began with the

Dutch (Grove 1996:142–3). This is the colonial root of conservation. I argue that to appreciate it we need to understand both the power of Western knowledge and the political economy of extractive colonialism. Extractive colonialism awakened a consciousness of the need for conservation that had already forgotten its roots in local understandings of the environment.

The emergence of the concept of "indigenous knowledge"

The international discourse of indigenous knowledge[6] began with an academic and activist effort to elevate the low status given to local knowledge and help people whose land rights were threatened. One of its pioneers was certainly Darrell Posey (1985), a British anthropologist who argued the case that the Kayapó of Brazil were actively managing their savanna ecology based on their indigenous knowledge, and that giving them land rights would ensure conservation (see Dove and Carpenter 2008:2–6). This academic work caught on with NGO activists and journalists who took it all over the developing world. It is sometimes referred to as the discourse of "indigeneity." It still carries Posey's original argument, that "indigenous" people know and manage their environment well, and thus should continue to be allowed to manage it by being given land rights. Establishing that a group has indigenous knowledge is always part of the activist effort to help local people gain land rights. Academics now more often analyze these activist efforts. Roy Ellen (Ellen, Parks, and Bicker 2000, and see Chapter 4) edited a recent example of this approach. Tania Li[7] contributed a chapter to that volume, in which she explores the implications of the fact that only the less apparently "indigenous" of two indigenous groups in Sulawesi, Indonesia, was able to claim that it possessed "indigenous knowledge." She has published on the same case elsewhere as well (Li 2000a, 2007).

In Indonesia, as in many other developing nations in the early 1990s, NGOs and local people were working together to establish rights. The discourses of indigeneity and indigenous knowledge were accruing power internationally, though that power was often not recognized at the national level. In any case, claiming "indigenous" status always involved documenting the possession of indigenous knowledge (Li 2000b:128). Li analyzes the effort to claim rights as a "field of power" characterized by diverse and complex agendas, "within which alliances may be formed, struggles waged, claims made and rights asserted (or denied)" (Li 2000b:121). There were internal and external aspects to that struggle.

In the 1990s the New Order Indonesian state refused to recognize the term "indigenous" (Li 2000a:149). Reviewing the positions of three government departments and two documents (prepared by government departments, donors, and NGOs), however, Li (2000b:124) concludes that the state is divided, with the Department of Social Affairs denying the existence of indigenous people, knowledge, or rights, while others (Environment Ministry, Ministry of Forests) argue for rights based on proven traditional or tribal environmental knowledge "as an adjunct to a conservation agenda." The conservation and biodiversity agenda seemed to offer an agenda upon which some parts of the state and NGOs could agree. The Lake Lindu area attracted NGO attention because the state was proposing to build a dam there. That is the background of the external aspect of the struggle for rights.

There are internal difficulties also. People accepting the "indigenous" label "must be ready and able to articulate their identity in terms of a set of characteristics recognized by the government agencies ... NGOs and by the media," which includes "the production of a reified body of 'indigenous environmental knowledge' from the pool of everyday local knowledge and practice" (Li 2000b:129). The Lindu people, although mission-educated, literate, fairly prosperous Christians who had been resettled near Lake Lindu by the Dutch, had a history of resource struggles with more-recent migrants to the area. These struggles informed their willingness to take on the work of resisting the state dam project, which threatened to resettle them again. They decided "to articulate (select, formulate and convey) a set of Lindu *adat* rules [traditional law] which *ought* to be acknowledged by outsiders, a process which in turn reworked the significance and substance, of Lindu knowledge and identity" (Li 2000b:130, author's italics). The Lindu are not passive in their articulation: they articulate with the discourse of indigenous knowledge, but also mention their Christianity and their adequate standard of living, positioning themselves as "ordinary rural people" with "special knowledge and an attachment to place" (Li 2000b:132–3).

Not all people who might qualify as indigenous can articulate with the indigenous discourse: the hill farming Lauje do not. The Lauje never had a historical need "to discover, constitute or record Lauje practices or traditional law (*adat*)"; "They do not articulate a sense of *adat* or tradition as something distinctive, autochthonous, locally derived or essential to Lauje identity" (Li 2000b:136–7). As Li (2000b:135) explains: "There has been no conjuncture, no context, site, event or encounter, in which the mountain Lauje have articulated a collective position as 'indigenous people'"; "The concept of 'indigenous people' has not taken hold in the Lauje hills because, under current conditions, it would not help people to make sense of their situation, nor would it help them to improve it."

The contrast between the Lindu and the Lauje allows Li (2000:142–3) to conclude that what constitutes "indigenous environmental knowledge" in Indonesia "has less to do with the innate qualities or wisdom of particular groups than with the power-laden processes through which knowledge and identity are formulated, communicated and brought to bear at particular conjunctures." We might go farther and say that because claiming indigenous status requires articulating with an external discourse, a group cannot actually be indigenous to claim that status. In other words, to claim to be indigenous, a group must stand outside itself and look back, self-consciously, at itself, from an external point of view (see Hirtz 2003). This process is power laden and only occurs at particular historical conjunctures.[8]

The documentation of indigenous environmental knowledge in the Indonesian context eerily foresees the integration of TEK in the Yukon described by Nadasdy. NGOs do the actual work of articulation in alliance with indigenous people (in the Yukon they are TEK researchers or First Nations peoples themselves or their representatives). The divided state, alongside donor agendas (and funding) offers "opportunities and constraints" that shape that work. Scientific knowledge does that shaping in the Yukon. In Indonesia, the NGO work includes preparing "documents, lists, rules and regulations as well as maps showing territorial boundaries and land-use zones, all of which help to make the idea of a community with customary practices and rights more visible to the state" (Li 2000b:128). In the Yukon, lists of species, numbers, and maps translate TEK into science. What NGOs produce in Indonesia is "a reified body of 'indigenous environmental knowledge' from the pool of everyday local knowledge and practice" (Li 2000b:129). In the Yukon a "way of life" is compartmentalized and distilled.

Purposeful integration of science and indigenous knowledge

In Paul Nadasdy's[9] piece, the colonial domination of indigenous peoples and the successful claiming of indigenous knowledge lie in the past; in the research present, a Canadian conservation project is intentionally integrating scientific and traditional First Nations' knowledge for environmental co-management. Post-colonial policy has seemingly embraced TEK. In reality, Nadasdy argues, "traditional ecological knowledge" has become a discourse with the power to be embraced in policy, but also the power to block understanding of what traditional knowledge really is.

In Nadasdy's ethnography, both the discourse of TEK and actual First Nations' knowledge exist. I would argue, and I think that Nadasdy is essentially arguing, that the discourse of TEK works against a successful integration of scientific and traditional knowledge for environmental co-management. We can see this failed integration as a particular sort of articulation—an articulation colored by the powerful discourses of science and TEK. The articulation described by Grove, in contrast, actually was an integration, though it was transformed by the powers of publication and colonialism.

What is actual First Nations' TEK? Nadasdy describes it, first, as not traditional in the sense of being static (Nadasdy 2003:120). Second, First Nations' knowledge does not separate environmental from human knowledge. Elders asked to share their environmental knowledge may talk about kinship or respect rather than animals or landscapes (Nadasdy 2003:121). "Stories, values, social relations, and practices" all contribute "substance and meaning to aboriginal peoples' relationship to the environment," such that they see animals as "sentient members of the social, moral, meaning-filled universe of the hunter and his or her family" (Nadasdy 2003:126).[10] Finally, First Nations people say that traditional knowledge "is not so much knowledge as it is a 'way of life'" (Nadasdy 2003:121). It is "one aspect of broad cultural processes that are embedded in complex networks of social relations, values, and practices" (Nadasdy 2003:121). One experienced hunter said that survival in the bush depends on knowledge of the environment as a whole: "*all* of the animals out there—how they behave, what they eat, how they interact with one another, and how they think" (Nadasdy 2003:124).

Nadasdy's description of the political problem with integration could have been theorized as the power of discourses, though he does not use Foucault in this way. The meanings of "traditional," "ecological," and especially "knowledge" are not shared by scientists and First Nations peoples, and "constrain people's thoughts and actions in significant ways" (Nadasdy 2003:120). They play a role "in structuring the way that people can act upon and think about TEK and its relation to science" (Nadasdy 2003:120). The integration is biased in favor of science, he writes, "by restricting the ways in which it is possible to talk (and think) about these issues" (Nadasdy 2003:119). Nadasdy briefly reviews the specific problems with "traditional" (which allows dismissal as inauthentic) and ecological or environmental (much as Cronon on wilderness does), before focusing on "knowledge," his main point. As he writes:

> That traditional knowledge might be used to rethink unexamined assumptions about how people should relate to the world around them, which unconsciously form the basis of scientific wildlife management itself, is a possibility that scientists and resource managers never entertain.
> *(Nadasdy 2003:122)*[11]

Foucault's concept of the power of discourses suggests that they cannot entertain this possibility.

Given the power of TEK to constrain the words, acts, and thoughts of scientists, how does the articulation occur? In other words, how does the discourse of TEK, working on scientists, shape the articulation between TEK and science? Nadasdy describes two processes: compartmentalization and distillation. Compartmentalization extracts information from its social context and divides it into scientific specialization (e.g., sheep TEK) (Nadasdy 2003:124–5). Distillation transforms that information into things like numbers and maps (according to TEK, how many sheep are located where). Together, these processes transform what is defined as "holistic, oral, qualitative, and intuitive" into TEK artifacts, which are "categorized, written, quantitative, and analytical" (Nadasdy 2003:129). Grove's Dutch governor was more respectful of his local physician, allowing him to set the structure for the *Hortus* as well as largely selecting the plants to be included.

TEK artifacts are essentially Latourian inscriptions; I argue that both compartmentalization and distillation transform the living understandings of First Nations' peoples into the immutable mobiles that science and management need. Living understandings, in their context and with all their complexity, are useless to scientific resource management until they are translated in this way. This translation has powerful effects, but it occurs without the awareness of TEK researchers. Nadasdy argues that TEK simply "must be expressed in forms that are compatible with existing institutions and processes of scientific resource management" (Nadasdy 2003:123). TEK researchers, he writes, "have no choice but to conform to the meanings scientists and resource managers assign to terms like 'knowledge' and 'traditional' when designing and conducting TEK research" (Nadasdy 2003:130). I would argue that meanings are not *assigned* in this way, and that conforming may also not be a question of *choice*. Scientists need sheep TEK because they are sheep scientists—knowledge shapes disciplinary specializations that shape the institutions of learning, research, and management within which they think. This is Foucault's insight. Numbers and maps have more power in science than descriptions of a way of life. This is Latour's insight.

For Nadasdy, what is critical is the fact that science is "incorporated into and legitimized by the institutions of state power" (Nadasdy 2003:138). This is true, but perhaps not critical. I argue that we have to go deeper to understand the failure of an integration shaped by the power of scientific knowledge. Science itself simply cannot fathom living ecological understandings.

Nadasdy uses Latour's concept of metrology to make one final point: that the integration of TEK into science is "a process that is extending the social and conceptual networks of scientific resource management into local communities" (Nadasdy 2003:141). He is arguing that a mandated integration of TEK and science is reinforcing science, and reinforcing the actual power that science has over First Nations' real natural resources. Nadasdy's solution is to give decision-making control of their territories to First Nations' peoples themselves (Nadasdy 2003:145). This may be unlikely, but it is the only real solution.

Nadasdy critiques a cutting-edge conservation project that mandates the integration of indigenous knowledge and science for the purposes of managing natural resources. He establishes the fact that indigenous knowledge and science cannot be integrated. Until scientists can respect sheep elders (and trade with bees), among other things, they are deaf to real, traditional but living, ecological knowledge. I tend to believe that this isn't any better for the sheep than for the First Nations peoples.

One final point: even though this project fails to actually integrate TEK and science, it does pull them together into an assemblage (see Chapter 14) that persists. I will refer back to this example of TEK in Chapter 15.

Conclusion

In this chapter we have examined the play of power between knowledges in a history and two ethnographies. In Grove's history, colonial botany actually meets local knowledge on more or less equal terms; this allows local knowledge to influence the structure as well as the contents of colonial botany, but it does not prevent its absorption and thus disappearance in time. The discourse of indigeneity, detailed in Li's ethnography, offers hope for land rights because it enshrines the idea that indigenous people are the best land managers; but it also carries the idea that those people must be authentically traditional, thus circumscribing the ways they can use those rights. It also forces local people to alter what their knowledge of their environment actually is. And it does not elevate indigenous knowledge to the status of scientific environmental knowledge, as we see in Nadasdy's ethnography. In the Yukon, indigenous knowledge and scientific knowledge are supposed to be equals informing co-management together. But this intent is entirely undermined by the seemingly neutral processes of compartmentalization and distillation.

Though none of these authors use Foucault's concept of the power of knowledge, they all could have. The science growing out of the articulation documented by Grove has forgotten its indigenous roots, and that forgetting testifies to the power of the idea of a Western science unlike and superior to indigenous knowledge. In Indonesia, when "indigenous knowledge" becomes a powerful discourse (especially among conservationists), groups with some indigenous qualities are able to ally with activists and perhaps tap into that power—though that requires them to distance themselves from their identity and knowledge. The mandated integration Nadasdy analyzes, similarly, fails because scientists and managers simply cannot think outside the powerful box of their science.

Other sorts of power are evident in these three ethnographies as well. Latour's work is helpful in understanding how van Reede's publication of essentially indigenous knowledge grew into an economic and political empire. In Sulawesi, NGO activists helped the Lindu translate their knowledge into the rules and maps that the state requires. In the Yukon, numbers and maps turn ways of life into TEK. In addition, Nadasdy uses Latour to understand how this transformed TEK has effects of power beyond that initial translation: over time First Nation minds, societies, and natural resources become more amenable to scientific management.

In all these readings, actual economic and political power also plays a role. In Grove and Nadasdy this power follows transformations in the realm of knowledge; in Li it provides the context in which a discourse becomes a small opportunity.

I will not discount the small opportunity that Li sees: indigenous knowledge should continue to be used as an activist tool for land rights. But it is clearly a very flawed tool, infected by the contradiction between requiring people to move away from lived knowledge to document their possession of "indigenous knowledge," and leaving those same people open to the judgment of lack of authenticity once they have done so.

Combining Tsing (from Chapter 7) and Nadasday's solutions offers a challenge to conservation. Nadasdy's solution is to give decision-making control of their territories to First Nations' peoples themselves, because science cannot hear their wisdom. Tsing's solution is respectful multi-cultural conversations, better frameworks for listening. Nadasdy clearly believes that First Nations' way of life is best for their environment. Tsing believes that "open-ended, semi-autonomous management relations" make it possible for people and forests to live together for a long time. These are the solutions that ethnography offers to replace government.

Notes

1. Tania Li (2000a:152), borrowed the term "articulation" from Stuart Hall (1996), to carry the dual meaning of "the process of rendering a collective identity, position, or set of interests explicit (articulate, comprehensible, distinct, accessible to an audience), and of conjoining (articulating) that position to definite political subjects." This meaning is relevant for the following sections on Li and Nadasdy, but not for Grove.
2. The opposition between Western/indigenous and global/local knowledges is not real, but part of a discourse itself, which is made differently in each of the following ethnographies.
3. Grove is a professor at the Centre for World Environmental History at the University of Sussex. See Grove 1995. In 2006, Grove suffered a very serious car accident in Australia, and has been severely incapacitated since that time.
4. The theory most useful to understand the importance of publications is Latour. Grove mentions Latour, and Nadasdy explicitly uses him. For more on Latour, see Chapter 7.
5. Grove notes that Latour's *Science in Action* is "helpful to those attempting histories of cultural constructions of nature," and refers to the capital of Malabar as "the colonial centre of calculation" (Grove 1996:127, 140).
6. Indigenous knowledge is typically imagined to be essentially different from Western scientific knowledge, but social scientists have argued that this is not true at all (e.g., Agrawal 1995).
7. Li is a professor in the Department of Anthropology at the University of Toronto. Li 2000a (149) grew out of an environmental politics seminar at Berkeley; Li 2000b grew out of a 1997 workshop on indigenous environmental knowledge at the University of Kent.
8. We should also note that the Western discourse imagines an authentic indigenous that is entirely unselfconscious. See Conklin and Graham 1995.
9. Nadasdy is an associate professor in the anthropology department of Cornell University.
10. This echoes Anna Tsing in Chapter 7.
11. This also echoes Tsing in Chapter 7.

References

Agrawal, Arun 1995 Dismantling the Divide between Indigenous and Scientific Knowledge. *Development and Change* 26:413–39.

Conklin, Beth A. and Laura R. Graham 1995 The Shifting Middle Ground: Amazonian Indians and Eco-Politics. *American Anthropologist* 97(4):695–710.

Dove, Michael R. and Carol Carpenter, eds. 2008 *Environmental Anthropology: A Historical Reader.* Malden, MA: Blackwell.

Ellen, Roy, Peter Parks, and Alan Bicker, eds. 2000 *Indigenous Environmental Knowledge and Its Transformations: Critical Anthropological Perspectives.* Amsterdam: Harwood.

Grove, Richard H. 1995 *Green Imperialism: Colonial Expansion, Tropical Island Edens and the Origins of Environmentalism 1600–1860.* Cambridge, MA: Cambridge University Press.

Grove, Richard H. 1996 Indigenous Knowledge and the Significance of South-West India for Portuguese and Dutch Constructions of Tropical Nature. *Modern Asian Studies* 30(1):121–43.

Hall, Stuart 1996 On Postmodernism and Articulation: An Interview with Stuart Hall. Lawrence Grossberg, ed. In *Stuart Hall: Critical Dialogues in Cultural Studies.* David Morley and Kuan-Hsing Chen, eds. Pp. 130–50. London: Routledge.

Hirtz, Frank 2003 It Takes Modern Means to be Traditional: On Recognizing Indigenous Cultural Communities in the Philippines. *Development and Change* 34(5):887–914.

Li, T.M. 2000a Articulating Indigenous Identity in Indonesia: Resource Politics and the Tribal Slot. *Comparative Studies in Society and History* 42(1):149–79.

Li, T.M. 2000b Locating Indigenous Environmental Knowledge in Indonesia. In *Indigenous Environmental Knowledge and Its Transformations: Critical Anthropological Perspectives.* Roy Ellen, Peter Parkes, and Alan Bicker, eds. Pp. 121–49. Amsterdam: Harwood Academic Publishers.

Li, T.M. 2007 *The Will to Improve: Governmentality, Development, and the Practice of Politics.* Durham, NC: Duke University Press.

Nadasdy, Paul 2003 *Hunters and Bureaucrats: Power, Knowledge, and Aboriginal–State Relations in the Southwest Yukon.* Vancouver, BC: University of British Columbia Press.

Posey, Darrell A. 1985 Indigenous Management of Tropical Forest Ecosystems: The Case of the Kayapó Indians of the Brazilian Amazon. *Agroforestry Systems* 3:139–58.

9

SUBJECT FORMATION IN FOUCAULT

Introduction

In 1983, near the end of his life, Foucault (1983:208–9) wrote that the subject, and not power, had been the goal of his work for the past 20 years. Subject formation provides another tool for the tool-box of power in conservation.

The subject in Foucault is a free subject. With the exception of coercive sovereignty, when Foucault is talking about power, he is talking about power relations between free subjects. So the subject's freedom is the key to non-coercive power in Foucault. And the sorts of power associated with subject formation—discipline and governmentality (and bio-power)—are not coercive. Discipline and governmentality act on and function through the subject's free and active participation, not by imposing a rule of law on the subject.

But the subject's freedom is not the freedom we imagine: the knowledges, practices, and technologies that create and define that freedom are not invented by the self: "They are patterns that he finds in his culture and which are proposed, suggested and imposed on him by his culture, his society and his social group" (Foucault 1988:11). In Foucault's (1986:68) terms, truth (discourses with the power of truth) is central to subject formation. Foucault (1988:19) notes that the freedom of the subject implies or assumes a relation of self to self. Free subjects care for themselves; they "self care." This requires a knowledge of self, which is defined by truth—meaning a discourse of power about what a self should be and how it should be attained (Foucault 1988:5). Subjects have agency—the freedom to act—but not autonomy—the freedom to set one's own laws.

Nevertheless subjects can and do resist subjectivities. Foucault (1983:213) believed that such a struggle was occurring in the "nowadays" of 1983, and also linked it to the Protestant Reformation. He believed that philosophy could contribute to the promotion of "new forms of subjectivity" (Foucault 1983:216). One can work to become aware of the knowledges and practices shaping one's subjectivity—through an analysis of history or ethnography—and thus perhaps choose different knowledges and practices to define one's subjectivity.

The government of the self by the self is always present in the literature on government. Foucault (2007:88) tells us that the "general problem of 'government'" emerges in the sixteenth century with respect to the problem of the government of oneself, of souls, of children, and of the state. This marks the break from sovereign power. The art of government literature spells out a series: one must first govern himself, then his family, his goods, his lands, and only then the state (Foucault 2007:94).

Different kinds of government form subjects differently. Near the end of his life, Foucault (1983:208) traces three modes of objectification that form subjects: "the modes of inquiry which try to give themselves the status of sciences"; the "dividing practices" in which "the subject is either divided inside himself or divided from others"; and "the way a human being turns him- or herself into a subject." Foucault (1983:212) also says that there are two meanings of the word subject: "subject to someone else by control and dependence, and tied to his own identity by a conscience or self-knowledge." Dividing practices are part of disciplinary government, which developed the arts of transforming people into subjects. Turning oneself into a subject is part of governmentality, which governs in part by relying on subjects to govern themselves.

The sciences are the first mode of subject formation. The birth of the human sciences in the beginning of the nineteenth century heralded modern subject creation. For the first time, people were the objects of knowledge, empirical entities to be known (Foucault 1970:344–87). The human sciences make the individual the object of analysis (e.g., in economics, the subject who labors is objectified). In other words, subjects are formed by knowledges that make the individual the object of analysis. Knowledges with the power of truth define our bodies, minds, and behaviors. "To care for self is to fit one's self out with these truths" (Foucault 1988:5). Disciplinary power and the human sciences have a particularly intimate and productive relationship.

Second, the subject is divided from others: sinner/redeemed, mad/sane, sick/healthy, prisoner/guard. I link this to disciplinary subject formation, in which states and a variety of institutions (e.g., schools, the military, factories, prisons) and sciences (e.g., medicine, psychiatry) see governing as shaping life.

Third is the way human beings turn themselves into subjects; individuals are "tied to [their] own identity by a conscience or self-knowledge" (Foucault 1983:208, 212). Governmental subject creation works by inducing individuals to observe, judge, and correct their own behavior (even their own thoughts), thereby conducting themselves towards some new purpose. This mode of subject formation is essential to governmentality.

Subject formation is being used in any conservation intervention that intends to make people into conservationists.

The disciplined subject

In "The Means of Correct Training," Foucault (1977:170) writes that "discipline 'makes' individuals," regarding individuals "both as objects and as instruments of its exercise." It makes individuals with three "simple instruments": hierarchical observation, normalizing judgment, and their combination in the examination (Foucault 1977:170).

In the first mechanism, hierarchical observation, subjects are made by "the spatial 'nesting' of hierarchized surveillance" in the architecture of military camps, working-class housing estates, hospitals, asylums, prisons, schools, workshops, and factories (Foucault 1977:171–2). This sort of architecture operates "to transform individuals" through "the disciplinary gaze" (Foucault 1977:172, 174). This surveillance creates the cooperative workers needed by capitalists (Foucault 1977:175). In conservation, surveillance is most often seen inside parks.

The second mechanism of disciplinary subject creation is normalizing judgment. Normalization works by comparing, differentiating, hierarchizing, homogenizing, and excluding. In other words, normalization compares individuals to each other (grades them), differentiates between them, ranks them (rewarding those with higher ranks), induces them to conform to a norm, and excludes (even punishes) individuals who do not conform. He notes that norms judge areas that laws have left untouched, like time, the body, and sexuality (Foucault 1977:178). Norms also judge areas relevant to conservation: big-game hunting, tree felling.

The third mechanism is the examination, which combines the techniques of surveillance and normalization in the "normalizing gaze" (Foucault 1977:184). The normalizing gaze dictates that its subjects must be visible; this visibility is fixed in "the file" (Foucault 1977:187, 190). Correlations between files determine averages and thus fix the norms against which individuals can be measured. Each individual becomes "a case," which means both an object of knowledge and of power; the individual is constituted as "effect and object of power, as effect and object of knowledge" (Foucault 1977:191, 192). The individual becomes the unit of knowledge and power, "the calculable man" (Foucault 1977:193). Power produces both "the individual and the knowledge that may be gained of him" (Foucault 1977:193–4).

Foucault traces a change within disciplinary power from the plague to the panopticon that occurs in the seventeenth and eighteenth centuries (Foucault 1977:205, 209). The panopticon was developed by the English social theorist Jeremy Benthan in the late eighteenth century. It is an architectural plan designed to allow all inmates of an institution (prisons, hospitals, schools, sanatoriums, and asylums) to be observed by a single watcher, without the inmates being able to tell whether or not they are being watched. Bentham described the panopticon as "a new mode of obtaining power of mind over mind, in a quantity hitherto without example" (Wikipedia n.d.). Foucault says that the individual is carefully fabricated in it; it insidiously objectifies those on whom it is applied (Foucault 1977:217, 220). He also notes that the panopticon increases both the docility and the utility of the individuals in it. It arranges power, but not for power itself: "its aim is to strengthen the social forces—to increase production, to develop the economy, spread education, raise the level of public morality; to increase and multiply" (Foucault 1977:207–8).

In a panopticon, inmates are caught up in a power situation of which they are themselves the bearers, because the surveillance is permanent in its effects (actual surveillance does not matter, nor does it matter who is watching); the inmate "inscribes in himself the power relation in which he simultaneously plays both roles," observed and observer (Foucault 1977:201–3). This is close to pastoral power, which by 1983 Foucault sees as being the source of the third mode of subject creation: the way human beings turn themselves into subjects. Pastoral power conceals a listener (the priest during confession); the panopticon conceals a watcher. In effect the panopticon both divides its inmates and forces them to internalize their observation.

The growth in disciplinary power and the formation of new knowledges "regularly reinforce one another in a circular process." The knowledges, all human sciences, include: clinical medicine, psychiatry, child psychology, educational psychology, and the rationalization of labor. Foucault (1977:224) describes this as a "double process": "an epistemological 'thaw' through a refinement of power relations; a multiplication of the effects of power through the formation and accumulation of new forms of knowledge." The flowering of power in certain local centers like child sexuality leads to the flowering of new knowledge, which in turn amplifies power.

Note that these knowledges correspond to disciplinary institutions that are still with us. We still discipline school children and prisoners, hospital patients, factory workers, and members of the armed forces, as well as people in many parks and buffer zones. We discipline those who are not yet capable of disciplining themselves, those who have demonstrated an inability to discipline themselves, or those who must be reduced to cogs in a machine for that machine to function efficiently: medicine, mental health, industry, or military.

Without the concept of the "population," the disciplines nevertheless governed multiplicities of people effectively: "The disciplines are the ensemble of minute technical inventions that made it possible to increase the useful size of multiplicities by decreasing the inconveniences of the power … which must control them" (Foucault 1977:220). Capitalism both helped and fed off of this process.[1] Foucault (1977:221) says that the accumulation of people and the accumulation of capital cannot be separated:

> It would not have been possible to solve the problem of the accumulation of men without the growth of an apparatus of production capable of both sustaining them and using them; conversely, the techniques that made the cumulative multiplicity of men useful accelerated the accumulation of capital.

All of these disciplinary mechanisms constituted the "dark side" of a legal system guaranteeing individual rights:

> The general juridical form that guaranteed a system of rights that were egalitarian in principle was supported by these tiny, everyday, physical mechanisms, by all those systems of micro-power that are essentially non-egalitarian and asymmetrical that we call the disciplines.
>
> *(Foucault 1977:222)*

In fact: "The real, corporal disciplines constituted the foundation of the formal juridical liberties" (Foucault 1977:222). The subject's legal liberties and the disciplines developed together, but in contradiction to each other. In governmentality, in contrast, the freedom of the subject is enlisted to govern individuals.

The subject of governmentality

Subject formation has a special relationship with governmentality. The practices of self and governmentality mutually presuppose each other (Dean 1994:163). In Foucault's terms, governmentality is the "encounter between the technologies of domination of others and those of the self" (Foucault 1997[1982]:225). Technologies of power "determine the conduct of individuals and submit them to certain ends or domination, an objectivizing of the subject" (Foucault 1997:225). Technologies of the self, in contrast:

> permit individuals to effect by their own means, or with the help of others, a certain number of operations on their own bodies and souls, thoughts, conduct, and ways of being, so as to transform themselves in order to attain a certain state of happiness, purity, wisdom, perfection, or immortality.
>
> *(Foucault 1997:225)*

In governmentality, there is no contradiction between the conduct of conduct and the freedom of subjects (as there was between the exercise of disciplinary powers and the legal framework of free individuals), because those free subjects are conducting themselves.

Technologies of the self enable what Foucault calls ethical subject formation, the government of the self by the self. The "others" who may help the ethical subject are nothing like the warden who may or may not be watching the inmate. They are the experts defined by the human sciences: the doctors, psychiatrists, psychologists, teachers, etc. Subjects increasingly seek them out in order to improve themselves, their "bodies and souls, thoughts, conduct, and ways of being."

Foucault argues that the governmental state looks after the welfare of the population, as we have seen, but it does not ignore individuals: it governs individualization; that is, it shapes individuality. As Foucault (1983:214) writes:

> I don't think that we should consider the "modern state" as an entity which was developed above individuals, ignoring what they are and even

> their very existence, but on the contrary as a very sophisticated structure, in which individuals can be integrated, under one condition: that this individuality would be shaped in a new form, and submitted to a set of very specific patterns.

The development of the knowledge of people focused on two roles, one concerning the population, especially the statistical social sciences; the other concerning the individual, especially psychology and psychiatry. As part of governmentality, subject formation works alongside the government of population. Governmentality is a "tricky combination" of the "totalization procedures" of governing a population, and "individualization techniques" that create governmental subjects (Foucault 1983:214). Government binds individuals conducted by others to the way they conduct themselves (Foucault 1993:203).

Historically, Foucault (1993, 1983) traces individualization techniques back to Christianity, to pastoral power and the self-verbalization of confession. In Christianity the purpose was adherence to religious commandments: congregants were expected to observe and attempt to correct their own behavior, and to verbalize lapses to the pastor as judge. As Foucault (1983:214) says, "This form of power cannot be exercised without knowing the inside of people's minds, without exploring their souls, without making them reveal their innermost secrets." The "individualization techniques" that originated as pastoral power came to constitute a "government of individualization" (Foucault 1983:212). He says: "We can see the state as a modern matrix of individualization, or a new form of pastoral power" (Foucault 1983:213, 215). Pastoral power is a form of power that "categorizes the individual, marks him by his own individuality, attaches him to his own identity, imposes a law of truth on him … a form of power which makes individuals subjects," ties them to their "own identity by a conscience or self-knowledge" (Foucault 1983:212). The modern state doesn't ignore the individual; it integrates individuals by shaping their individuality in a new form, which is characterized by self-knowledge.

Foucault (1983:216) wants us to refuse what we are, liberate ourselves from the type of individualization that is linked to the state. A "new economy of power relations" is his term here for what we need, what philosophers must theorize, "not to discover what we are, but to refuse what we are" (Foucault 1983:210, 216). As he says, "We have to promote new forms of subjectivity" (Foucault 1983:216). He suggests we analyze governmentality through analyzing resistances to it, especially contemporary resistances that are "struggles which question the status of the individual," struggles against the "government of individualization," and struggles that question the regime of knowledge that governs individuals (Foucault 1983:211–12). In an interview in 1984 Foucault (1988:18, author's italics) said:

> The problem is not of trying to dissolve [relations of power] in the utopia of a perfectly transparent communication, but to give one's self

the rules of law, the techniques of management, and also the ethics, the *ethos*, the practice of self, which would allow these games of power to be played with a minimum of domination.

Conclusion

Subjects are constructed by knowledges that study them and use them as units of analysis. The human sciences in particular make the individual the object of analysis. Subjects are cultivated by disciplinary mechanisms like surveillance, pressure to conform to norms, the maintaining of files and cases on individuals, and by architectures that separate them and manipulate them to observe themselves. Finally, subjects are created by the very concept of individuality, by encouraging people to know and improve themselves.

There are certainly projects that intend to create conserving subjects. But subject creation in Foucault is actually deeper than this: it is part of all the tools we bring to working with people. A forest council is fairly benign, but if we add different roles and rules within it, we are cultivating a certain sort of political subject; if we require dues, we are basing it on the rational economic individual. If we require guards to be hired and paid, we are doing these things plus creating the possibility of surveillance. If we require the punishment of forest interlopers, we are punishing the abnormal, and probably creating a file too. We are treating conservation like a norm, and on top of that treating the economic person as a norm, when all we wanted to do was to preserve the forest or the wildlife.

Note

1. Class domination also traverses discipline; the bourgeoisie was becoming the politically dominant class in the eighteenth century (Foucault 1977:222–3).

References

Dean, Mitchell 1994 "A Social Structure of Many Souls": Moral Regulation, Government, and Self-Formation. *Canadian Journal of Sociology* 19(2):145–68.

Foucault, Michel 1970[1966] *The Order of Things: An Archaeology of the Human Sciences.* New York: Vintage Books.

Foucault, Michel 1977[1975] *Discipline and Punish: The Birth of the Prison.* Alan Sheridan, trans. New York: Vintage Books.

Foucault, Michel 1983 Afterword: The Subject and Power. In *Michel Foucault: Beyond Structuralism and Hermeneutics.* 2nd edition. Hubert L. Dreyfus and Paul Rabinow, eds. Pp. 208–26. Chicago: The University of Chicago Press.

Foucault, Michel 1986[1984] *The Care of the Self,* Volume 3 of *The History of Sexuality.* Robert Hurley, trans. New York: Vintage Books.

Foucault, Michel 1988 The Ethic of Care for the Self as a Practice of Freedom: An Interview with Michel Foucault on January 20, 1984. In *The Final Foucault.* James Bernauer and David Rasmussen, eds. Cambridge, MA: The MIT Press.

Foucault, Michel 1993[1980] About the Beginning of the Hermeneutics of the Self: Two Lectures at Dartmouth. *Political Theory* 21(2):198–227.

Foucault, Michel 1997[1982] *Ethics: Subjectivity and Truth. Essential Works of Foucault: 1954–1984 Volume I.* Paul Rabinow, ed. Robert Hurley and others, trans. New York: The New Press.

Foucault, Michel 2007 *Michel Foucault: Security, Territory, Population: Lectures at the College de France, 1977–1978.* Michel Senellart, ed. Graham Burchell, trans. New York: Picador.

Wikipedia: The Free Library n.d. Panopticon. https://en.wikipedia.org/wiki/Panopticon accessed 10/10/2018.

10
SUBJECT FORMATION IN ETHNOGRAPHIES

Introduction

I have already examined one ethnography of conservation subject formation: the Nuaulu of Indonesia in Chapter 4. In that case Nuaulu subjects cultivated themselves as conservationists, primarily due to seeing the effects of degradation. Richard Grove similarly argued that knowledge of plants together with the visible effects of their extraction led the Dutch colonial government towards the idea and policy of conservation (see Chapter 8). The ethnographies that follow, however, all involve purposeful efforts to cultivate others as subjects. The methods of subject formation in the ethnographies, and in conservation in general, are a mixture of discipline and governmentality.

The first example concerns the role of knowledge in subject formation. The geographer Bruce Braun examines the impact of the birth of the science of geology and the first geological survey in Canada. Essentially, a new science redefines how territory is seen and governed, and the then-colonial state follows deliberate strategies to cultivate new geological subjects. This adds to Foucault's work on subject formation engendered by the advent of the human sciences. The second ethnography, by the anthropologist Tania Li, analyzes projects in Sulawesi, Indonesia, that were designed to cultivate conservation subjects, but instead educate a resistance to conservation and the invasion of a park. Li argues that subject formation is a fragile process. The final ethnography, by the political scientist Arun Agrawal, examines forest councils in northwest India that are successfully creating conservation subjects. He argues that practices are the key to subject formation, in contrast to Braun's emphasis on knowledge.

All of these authors use Foucault; Braun and Agrawal also use Latour.[1] All of these ethnographies are based on field research. In Braun's case this was his doctoral dissertation fieldwork. Li has visited the Sulawesi hills eight times, beginning in 1990. Agrawal visited Kumaon, India, three times, collecting interview and survey data twice.

This set of ethnographies stimulates us to think about how knowledge and practices contribute to subject formation, and about the potential for inducing conservation-mindedness versus the dangers of sparking a resistance. Braun's work is also a good example of the way Marxist and Foucauldian concepts can be used together.

Conservation policies and projects don't usually set out to "cultivate conservation subjects." Rather, they try to "incentivize" conservation in subjects viewed as deficient vis-à-vis conservation, or even degrading of the environment. Some policies simply aim to educate ignorant subjects: any conservation project with a training or educational component may be trying to cultivate conservation subjects with science. The first ethnography is relevant to these policies. Other policies are essentially conducting conduct, arranging things to "incentivize" the desired conservation behavior. Conservation projects creating new financial flows, from, for example, tourism or the sale of handicrafts, imagine that they are rewarding conservation behavior like respecting park boundaries and rules. These projects are also trying to cultivate conservation subjects, but they rarely do the math comparing what the park has taken away with what they are offering. Free subjects are free to weigh what they have lost against what they are being offered. The second ethnography is relevant to these policies. Policies that give new use rights to land, like India's forest councils, are hoping that having some "ownership" (which is often not literal ownership of land, but on the contrary implies that the subjects have to bear and pay for the responsibility of conservation) may lead to more conserving behaviors. Political and economic requirements to build management capacity are typically included—which suggests that subjects are seen as deficient in many ways. The third ethnography is relevant for these projects.

Conservation policies and projects are generally using the methods of governmentality to cultivate conservation subjects—but these methods are applied to subjects that are imagined as deeply deficient, like disciplinary subjects. This is problematic in many ways. Most seriously, conservation policies are assuming that their scientific truths (which may not actually be cutting-edge science), their political structures and processes, and their economic system are superior to local knowledge, politics, and economics—without even knowing what local knowledge, politics, and economics are. But above and beyond the deep inequity of this policy position is the reality that positive, incentivizing methods may not work on subjects treated as deficient.

Sovereignty may also be present in the history or context of the conservation intervention, and it will affect subject formation. It exists in the context of the second ethnography considered here, contributing to project failure. In the third ethnographic example it exists in history, generating distrust of the forest service.

Science and subject formation

Bruce Braun[2] focuses on the emergence of a new formation of power/knowledge (a new discourse) in late nineteenth century Canada: the geological survey of Canada (GSC).[3] He argues that three things—science, governmentality (using Foucault), and capitalist production (using Marxist geographers)—are "tangled together in knots of intertwined practices and objects" (Braun 2000:39–40). In essence, Braun is adding earth science to Foucault (who focused on the human sciences), and then adding the extraction of a newly discovered natural resource to that.

Braun (2000:26, author's italics) gives most weight to the science piece, saying that "practices of *representation* ... are both central to and necessary for nature's *material* transformation" for capitalist production. Nature became legible in a new way: "What it meant to 'see' and what was 'there to be seen' in specific landscapes changed dramatically" (Braun 2000:20). He is arguing that geology created new domains for economic and political rationality (Braun 2000:24). Using Latour to suggest that epistemic shifts in knowledge are created through cycles of accumulation, he argues that the transformation of knowledge preceded rule, or that the new geological survey preceded the governmentality of "vertical territory" (Braun 2000:19).

The survey and the new legibility of vertical territory it created also meant that the state took an interest in devising new instruments and tactics to compel people to use the new information "in an improving direction," to "optimize use of Canada's 'geological spaces'" (Braun 2000:27, 33). Citizens needed to be "remade" into "'geological subjects'" with eyes "trained to see the inner architecture of the earth" (Braun 2000:28–9). The instruments and tactics in the latter half of the nineteenth century included GSC publications and a library, the distribution of mineral kits and maps to schools, and GSC and provincial museums whose visitors were "invited to imagine the country in terms of its vertical structure" (Braun 2000:31). This dissemination of the new science is clearly governmental subject formation.

The legibility of nature, capitalist production, and subject formation were not autonomous or stable. For example, the regulation of property, especially mining rights, was a provincial, not a central government function—but the provinces had no surveys (Braun 2000:32). Provinces made their own mining laws and gathered their own statistics (Braun 2000:34–6). Essentially, Braun (2000:35) argues, provinces were attempting to adjust their governing of the conduct of individuals as their understandings of provincial geology and its commodification shifted. Governmentality, geology, and capitalist production were all changing; they were also inextricably interwoven (Braun 2000:39).

In his 2002 book, Braun adds an important dimension to the use of knowledge to cultivate subjects, albeit without referring to subject formation. First, Braun (2002:46, author's italics) sees geology as "constituting" landscape: "'legibility' is not something *in* nature awaiting discovery by the disinterested observer; it is achieved through historically situated representational practices." The surveys were not simply the objective documentation of a pre-given landscape; they "constituted" the landscape as an intelligible domain (Braun 2002:47). The surveys "constructed spaces of visibility (and spaces of invisibility) that were at once both partial and of immense social, ecological, and political consequence"; "at issue are not questions of instrumentality and interests, but of enframing and power" (Braun 2002:47). The powerful framing in the science of geology and in the surveys is being passed on to the new geological subjects they are creating.

The best example of this is the fact that the surveys divided west coast Canadian landscapes into "discrete knowledge domains," extracting specimens "not only from their organic or ecological relations with each other, but also from their place in other people's economies, histories, social and symbolic systems" (Braun 2002:48, quoting Mary Louise Pratt 1992:31). In a brilliant analysis of Dawson's[4] journals and photographs, Braun (2002:52–4) reveals a "double vision": "on the left primeval nature, on the right Native culture." In addition, "Geology, as a way of knowing, served to reinforce for Europeans their sense of difference from, and superiority to, premodern cultures" (Braun 2002:59). Furthermore, "as a people without Geology," Natives "had no legitimate claim to the resources of the land" (Braun 2002:59). The surveys "saw landscapes not in terms of their contemporary cultural geographies, but in terms of a future national development" (Braun 2002:48). Both ecology and First Nations peoples are excluded from geological knowledge and severed from Canada's future, and First Nations peoples are judged as inferior and stripped of resource rights. Braun is arguing that this vision is inseparable from the science of geology that was being taught in schools and displayed in museums.

Braun (2002:59, 57) concludes that:

> through its displacement of local landscapes into global orders [stratigraphical geology in the late 1700s and early 1800s made local landscapes legible as part of a larger global order], its erasure of human occupants, and its privileging of European science over other ways of seeing, geology served to annex new territories not only into domains of imperial science, but also into imperial forms of political and economic rationality.

These imperial forms of rationality were embedded in the new science of geology that was shaping new geological subjects. I would argue that they are also embedded in conservation.

Problems with conservation subject formation

Tania Murray Li[5] analyzes two integrated conservation and development projects with subject formation goals, managed by the Asian Development Bank (ADB) and CARE, in the area surrounding Lore Lindu National Park in Sulawesi, Indonesia.[6]

Li argues that interventions like integrated conservation and development demonstrate a "will to improve." The will to improve in this case is the "earnest desire" to achieve the benevolent goal of integrated conservation and development. She criticizes (as I do in Chapter 1) "the rush to identify hidden motives of profit or domination" because it "narrows analysis unnecessarily, making much of what happens in the name of improvement obscure" (Li 2007:9). Essentially, the will to improve constitutes Foucault's governmentality, conducting the conduct of a population for its own welfare.

The will to improve is implemented through two processes: problematization, that is, "identifying deficiencies that need to be rectified," and "rendering technical" (Li 2007:7).[7] Rendering technical draws on "an assemblage of techniques and calculations," in this case labeled integrated conservation and development, to identify "an arena of intervention," to bound it, to dissect it, and to devise "corrective measures to produce desirable results" (Li 2007:123). In other words:

> To render a set of processes technical and improvable an area of intervention must be bounded, mapped, characterized, and documented; the relevant forces and relations must be identified; and a narrative must be devised connecting the proposed intervention to the problem it will solve.
> *(Li 2007:126)*

The identified problem may be defined by the existence of an available solution (Li 2007:7). Problem and solution exclude political-economic problems that support systemic inequalities (Li 2007:11). Rendering technical also constitutes the boundary between those with deficiencies and those with expertise (Li 2007:7). Li views these processes as "a project, not a secure accomplishment" (Li 2007:10).

Li demonstrates this process by comparing research design documents with the final project plan. Design documents reflected thorough research on political-economic processes like the immigration of Bugis migrants attracted by the new cash crop, cacao, and the real cost to villagers of giving up park land, where many were growing coffee (Li 2007:124). Project plans, in contrast, "reposed political-economic causes of poverty and injustice," highlighting "only those problems for which a technical solution could in fact be proposed"; this excluded most of the collected data (Li 2007:126). As Li writes, "The proposition that development benefits would offset losses and compensate villagers had no empirical basis" (Li 2007:130). The plans proposed that improved farming for those with land, and income-generating projects for the landless, would protect the park (Li 2007:127).

The projects included efforts to cultivate the "will to conserve" in the local population. The ADB project, for example:

> proposed to create a new collective subject, a community that would assess, plan, reach consensus, and think of population and natural resources as

> entities to be managed. The proposed technique for creating this subject was to guide villagers through a carefully crafted sequence of activities: participatory assessment of community resources, problem analysis, preparation of development proposals, application for funding … monitoring, and evaluation of outcomes.
>
> *(Li 2007:132)*

The project population was problematized as deficient in conservation-mindedness, and the solution was rendered technical, drawing on a participatory assemblage already linked to "community." The plan offered project benefits to households or communities—development—in return for the signing of conservation agreements committing them to obey park rules. Benefits were livelihood improvements, essentially biochemical inputs (sacks of fertilizer and pesticide—Li notes how odd this is in a conservation project) and improved seeds for farming, and funding for infrastructure like roads, irrigation, and flood control (Li 2007:128–31). The ADB plan also included increased park enforcement and education of villagers and officials about the importance of conservation.

The CARE project also intended "to produce new subjects who would adopt ecofriendly practices on their own land and respect park rules," primarily "to devise regulations committing them to police each other and report on individuals who infringed park rules" (Li 2007:139). It promoted fruit trees and sustainable farming techniques like bench terraces, shade trees, alley cropping, green manures, and integrated pest management.

These efforts to cultivate subjects occurred in a context defined by sovereignty-like exclusion and enforcement. As we saw in Chapter 6, Li considers this to be in contradiction with the governmentality of the projects. People were excluded from Lore Lindu National Park in 1993, but many still had crops inside the park, and others were landless after exclusion. The new projects, in fact, increased the enforcement of exclusion from the park (Li 2007:131). Examples of sovereignty directly linked to the integrated conservation and development projects include destruction of crops and homes and arrests of villagers.

Both projects failed. The ADB agricultural interventions did not deliver "even modest gains" among farmers with land, and were useless for the landless. The infrastructure interventions were "monstrously corrupt," characterized by waste and incompetence (Li 2007:135–6). Five years into the project "no community conservation agreements had been developed or signed" (Li 2007:136). CARE's sustainable techniques were simply not adopted because they did not pay. Some interventions were captured by village elites, who were able to expand their land outside the park—but also inside it. The landless, who CARE saw as most likely to defy park rules, were not helped at all, because CARE couldn't identify any livelihood activities for them. (Li 2007:137–8). The results of phase one of this project were that two villages prepared regulations, and one reported an infraction (Li 2007:139). Essentially farmers were not interested in the development benefits either project offered, so the conservation didn't happen. It goes without saying that conservation subjects were not created by either of these projects.

The efforts to exclude political-economic processes and cultivate subjects failed. The projects had an unintended effect: the education of a resistance, of a newly modern civil society. In Li's analysis the failures of the projects inform new more "critical positions," alongside increasingly coercive park exclusion, growing landlessness, increased freedom to mobilize following the national coup in 1998, and a growing local NGO movement (Li 2007:142). The projects "stimulated new practices and the formation of critical communities in ways the planners certainly did not intend"; village meetings and discussions with experts "had a perverse effect: they clarified to the villagers that this set of experts could not help them solve their problems" (Li 2007:151). Increasingly violent clashes between villagers and park guards stopped only when one village, working with an NGO, persuaded the director of the park to recognize them as an indigenous community (Li 2007:147).[8] This opened the floodgates, in a movement that led to the occupation of Dongi-Dongi Valley inside the park, by a group calling itself Free Farmers Forum (Li 2007:150). CARE's midterm evaluation admits that "CARE appears, unwittingly, to have been an important facilitator of the Dongi-Dongi land occupation" (Li 2007:151). I would argue that the intention to cultivate conservation subjects in return for development contributed in two ways: first, by cultivating subjects, encouraging communities in village forums to talk about their problems, and second by setting up expectations that villager losses to conservation would be offset or balanced by development.

The Free Farmers Forum can be seen as the core of Li's book. As she says in its introduction, she is interested by how structural inequalities are excluded from improvement, but is also interested in:

> the "switch" in the opposite direction: in the conditions under which expert discourse is punctured by a challenge it cannot contain; moments when the targets of expert schemes reveal, in word or deed, their own critical analysis of the problems that confront them.
>
> *(Li 2007:11)*

It is the forum that showed Li, or at least confirmed to her, that government is always fragile, and subjects always unruly. She theorizes a relation of "permanent provocation"[9] between the practice of government, improvement rendered technical, and the practice of politics, the expression of a critical challenge (Li 2007:11–12).

Political-economic power played a role in all of this. There are many small-scale examples of this in the ethnography, like the capture of CARE's agricultural interventions by village elites. Li argues that such interests are part of development but not its "master term" (Li 2007:9). On a much larger scale, Li argues that "capitalism and improvement are locked in an awkward embrace" (Li 2007:21). Economic growth impoverishes some, and this impoverishment must be seen to be managed. Improving interventions give the appearance of addressing the "fallout from capitalism's advance" (Li 2007:21). But this appearance of management can fail, as it did around Lore Lindu Park—and then people may take things into their own hands.

A successful case of conservation subject formation

Arun Agrawal[10] makes extensive use of Foucault's ideas about government, governmentality, and subject formation—but he wants to ground theory in evidence, to "address subject formation concretely rather than abstractly" (Agrawal 2005a:167, 2005b:198). His findings demonstrate that people in villages with forests councils "had begun to view their and others' actions in forests in a way that valorized protection of trees and economy in the use of forest products" (Agrawal 2005a:173). Agrawal also spells out the implications of his research for policy: "Certain forms of environmental enforcement are associated with greater commitment to environmental conservation, higher levels of local involvement, and the generation of environmental subjectivities" (Agrawal 2005a:177).

Agrawal (2005a:180) argues that environmental subjects[11] in the villages of Kumaon, India, are formed by the working together of "the technologies of power that form subjects and encourage them to define themselves in particular ways and the technologies of self that individuals apply to themselves to transform their own conditions." "Technologies of power" began with the 1931 Forest Council Rules, in which the colonial government replaced fortress conservation of forests with community-based forest councils (Agrawal 2005a:178). The 1931 rules "tried to define, situate, and fix localities into a particular structural position—accomplices in conservation" (Agrawal 2005b:103). Agrawal theorizes this using concepts of "government at a distance," based on Latour's "action at a distance," and his own term, "intimate government." Government at a distance "presupposes centers of calculation, constant oversight, continuous collection of information, unceasing crunching of numbers, and the imposition of intellectual dominance through expertise" (Agrawal 2005a:178, citing Miller and Rose 1990:9–19). Intimate government, in contrast, "works by dispersing rule, scattering involvement in government more widely, and encouraging careful reckoning of environmental practices and their consequences among Kumaon's residents" (Agrawal 2005a:178). "Practice and sociality [community] rather than expertise form the basis of intimate government" (Agrawal 2005a:179). The forest law of 1931 supplemented government at a distance with intimate government (Agrawal 2005a:178). The Forest Department conceded the grand project of extensive control over forests only in return for a surer means of "intimate regulation" (Agrawal 2005b:91). The emergence of environmental subjects stems from both sorts of government: it "has been as much a consequence of processes marked by government at a distance as it has been about intimate government" (Agrawal 2005b:199).

In 1931, the Revenue Department, not the Forest Department, was given control over the formation of forest councils, and the councils' authority was limited, especially where state interests existed. A village petitioned the Revenue Department for permission to form a council (Agrawal 1995b:116). Villagers were given responsibility for conserving forests in exchange for harvesting subsistence products, while the Forest Department retained the benefits of commerce in timber and resin from villagers' forests (Agrawal 2005b:102). After a 1976 revision to the rules, councils had to get Forest Department approval to harvest more than one tree a year (Agrawal 2005b:118). The councils do the work of the Forest Department, guarding the forest and paying the guards (Agrawal 2005b:8). Another effect of the forest council structure was that it "dims the possibilities of protest and rebellion" (Agrawal 1995b:120). Forest councils are a nice example of governmentality. They give people something—in this case some rights to forests taken away during the colonial period—but specify conditions that are intended to induce the people to govern their own behavior vis-à-vis the forest—the councils and the rules.

Agrawal adds to technologies of power the "technologies of self": formed as subjects by the technologies of power, the people of Kumaon take over, transforming themselves. "Technologies of self" involved a shift in perceptions of personal interest that led some residents to monitor their own environmental practices and consequences, and to "care about" the environment. Agrawal (2005a:165) sees this shift in perceptions of interest and caring, what Foucault would see as self conduct, in terms of the imagination.

Practice is the key or crucial link for Agrawal between "power and imagination, between structure and subjectivity" (2005a:180, 162–3, 166, 176). His research demonstrates that participation in specific regulatory practices of government is most important: monitoring the forest or funding monitoring. Participation in these practices induces individuals to transform their own environmental values. This works because people "arrive at new conceptions of what is in their interest": subjectivity is a "palimpsest," a slate still bearing traces of past use, "on which involvement in institutionalized practices inscribes new and sometimes conflicting understandings of what is in one's interest over and over again" (Agrawal 2005a:166, 178). As Agrawal concludes, "Practices of specific subjects are the location where relationships between institutions and power/knowledge and imagination and subjectivity come together. They are a basic mechanism on which subject formation rests" (Agrawal 1995b:221).

There is another critical component of the imagination alongside shifting perceptions of interests. Agrawal (2005a:179, 2005b:197) coins the term "imagined autonomy" for a crucial part of this process. In Kumaon, communities develop "imagined autonomy" from the Forest Department: "stemming from precisely the practices of conservation encouraged by state officials," and using the language and idioms of protection that state officials use, communities govern their forests "in pursuit of goals that they imagine as their own" (Agrawal 2005a:179).[12]

Although the state is governing at a distance, it is doing so in ways "that local residents believe to be defined locally" (Agrawal 2005b:197). This ambiguity parallels the one Agrawal (2005a:162) points out between "the meanings of subject as agent or subordinate": Kumaon communities are subordinate to the Forest Department, but take on its purposes as their own.

In sum, Agrawal traces a process from participation in the practices of government through forest councils, through imagination, which combines a redefinition of a subject's interests together with imagined autonomy from the state, to the cultivation of new environmental subjects (Agrawal 2005a:173).

Ultimately, Agrawal is positive about forest councils: he wants to keep regulation in the community and he believes that it creates conserving subjects. This is a challenge to the critique inherent in Foucault and many who use Foucault. Agrawal's focus is on *how* new subjectivities formed. This focus allows him to use ethnography to ground theory, and to come to a measured conclusion about the effects of government and subject formation in Kumaon.

Agrawal differentiates between his analysis of subject formation, stimulated by Foucault, and political ecology. "The subject," he writes, "is always already present in political-ecological writings" (Agrawal 2005b:211). But, he argues:

> The relationships of subjects to the environment … need to be examined in their emergence, not simply taken as part of a larger politics by pre-existing interests. How environments, and the history of practices in relation to the environment, transform actors and interests is an enormously interesting and complex question.
>
> *(Agrawal 2005b:211)*

One cannot pursue this question without giving up "the concept of subjects and interests that are always already given by their social-structural locations" (2005b:211). We need to examine *how* subjects are made. Agrawal (2005a:165–6) is similarly critical of some of the governmentality literature on subject formation, saying that "much of the vast secondary literature on … governmentality, in contrast, defers a consideration of how subjects make themselves, focusing primarily on technologies of power aimed at objectifying individuals." In particular, he continues, "writings in the field of development and environmental conservation, even when influenced by Foucault … have been relatively inattentive to the variable ways in which self-formation takes place and how it may be shaped by involvement in different forms of practice" (Agrawal 2005a:166).

Conclusion

Braun, Li, and Agrawal provide ethnographies of subject formation. In Braun, new knowledge produced by the GSC was brought to the public through publications, museums, and schools as part of a strategy by national and provincial governments to cultivate geological subjects. In Li the ADB and CARE planned to produce conservation subjects by, first, rewarding conservation with development benefits, and, second, through a variety of participatory techniques, including meetings with experts. In Agrawal, participation in the practices of government in the form of forest councils responsible for guarding forests created environmental subjects.

Braun emphasizes knowledge and Agrawal practices. Li's projects had some components of both. The educational component of the projects Li analyzes, delivered as trainings rather than through publications, museums, and schools, simply failed.

Braun demonstrates that the knowledge of geology is not politically or economically or even ecologically innocent: it encoded separations between minerals, ecologies, and First Nations peoples, and was oriented towards capitalist extraction. Braun is not interested in the question of whether or not geological subjects were actually formed, so we don't know if these insidious aspects shaped subjects along with the science, but Braun suggests that they did. This should alert conservationists to examine our own knowledges for similar taints. Do we separate endangered species from their ecologies? Do we translate their value into economic terms? Do we pass these separations and translations along in trainings?

Agrawal's practices, those that actually formed environmental subjects, were the practices of government, while Li's were essentially the practices of participatory development. They do seem to have cultivated subjects, but subjects who realized that their problems were not going to be solved by these experts or these projects, not conservation subjects. The shift in perceptions of interest and imagined autonomy that Agrawal found in Kumaon do not occur. This suggests that the practices of government, *being given responsibility for an environmental domain* as well as benefits from it or to offset the costs of conservation, may produce conservation subjects. Participatory exercises create subjects who can think critically about whether their own interests are served by park rules.

Braun appropriately introduces Marxist theory into his analysis, because the intent of the geological survey, to optimize the use of vertical territory, is capitalist and extractive. Geology broadened the appreciation of the natural territory by Canada's colonial government, but it also provided a newly enlarged arena for power and capitalist extraction. Geological subjects were part of this project. The bifurcation of nature from First Nations people was also inherent in this project, and delivered to subjects along with the science of geology. "Native" settlements were not allowed to overlap with mineral riches; the fact that they were presumed to not know geology meant that mineral wealth had to be extracted from them. Though Braun does not trace the effects of this new knowledge on the environment, we can guess that it was not conserving.

Li's analysis traces a relation between capitalism and improvement that is also Marxist, but based on the work of the economic historian Karl Polanyi. She argues, as we have seen, that "capitalism and improvement are locked in an awkward embrace," because the unequal effects of capitalism, "fallout from capitalism's advance," must be addressed by improvement—but rendering technical excludes them. Rendering technical is bound to fail in contexts already defined by political-economic inequality. Proposing to improve agriculture in a project area where a quarter of the population is landless demonstrates the inability of improvement to address the effects of capitalism. In projects that "integrate" improvement with conservation, conservation is bound to be threatened, as it was in this case.

Agrawal goes further into subject formation than Braun or Li, particularly shedding light on how forest councils created not just environmental but self-governing subjects. Forest council participation can lead to a redefinition of self-interest and to "imagined autonomy," to subjects who have embraced forest department goals as their own and subordinates who imagine themselves as agents. Both require imagination. This delicate process simply never occurs in Li's ethnography. The dramatic failure of integrated conservation and development projects educates not conserving subjects, but resisting ones; self-interests are strengthened rather than redefined; real autonomy overtakes a poor effort to subordinate.

Notes

1. Li cites Agrawal, and read a version of Agrawal 2005a.
2. In at least one publication his name is Willems-Braun. He is a professor of geography at the University of Minnesota. His doctorate is from the University of British Columbia, and he did a post-doc at Berkeley.
3. The GSC was established in 1841, and is one of the first government organizations in Canada, and its oldest scientific agency (Wikipedia).
4. George Mercer Dawson was a nineteenth century Canadian geologist and surveyor, who joined the GSC in 1875.
5. Tania Murray Li is a professor of anthropology and Canada Research Chair in the Political-Economy and Culture of Asia at the University of Toronto. She has two books that are essential reading for environmental anthropologists: *The Will to Improve* (2007) and *Land's End* (2014).
6. The projects were implemented in the late1990s. The Indonesian state was democratized in 1998. The projects and responses that Li describes thus bridge the 1998 Indonesian coup.
7. Li adapts the phrase "rendering technical" from Rose 1999:33 (Li 2007:286n14).
8. As we saw in Chapter 8, communities in the Lake Lindu area had attempted to use the discourse of indigeneity to block state plans to build a dam.
9. She borrows this phrase from Foucault (1983:222).
10. Agrawal is a political scientist and professor at the School of Natural Resources and Environment, University of Michigan. He has strong ties to anthropology: he is married to the anthropologist Rebecca Hardin; this particular article was published in an anthropology journal, and the anthropologist Tania Li commented on a version of it.
11. Agrawal (2005b:165) defines subjects as actors or agents who are also subordinated.
12. Agrawal contrasts Benedict Anderson's imagination (a ruling group can bind a country into an imagined community) to James Scott's resistance (resisting subjects can protect their consciousness), placing Kumaon in between.

References

Agrawal, Arun 2005a Environmentality: Community, Intimate Government, and the Making of Environmental Subjects in Kumaon, India. *Current Anthropology* 46(2): 161–90.

Agrawal, Arun 2005b *Environmentality: Technologies of Government and the Making of Subjects.* Durham, NC: Duke University Press.

Braun, Bruce 2000 Producing Vertical Territory: Geology and Governmentality in Late Victorian Canada. *Ecumene* 7(1): 7–46.

Braun, Bruce 2002 *The Intemperate Rainforest: Nature, Culture, and Power on Canada's West Coast.* Minneapolis: University of Minnesota Press.

Foucault, Michel 1983 Afterword: The Subject and Power. In *Michel Foucault: Beyond Structuralism and Hermeneutics*. 2nd edition. Hubert L. Dreyfus and Paul Rabinow, eds. Pp. 208–26. Chicago: The University of Chicago Press.

Li, Tania Murray 2007 *The Will to Improve*. Durham, NC: Duke University Press.

Rose, Nikolas. 1999 *Powers of Freedom: Reframing Political Thought.* Cambridge, MA: Cambridge University Press.

Wikipedia: The Free Library n.d. Geological Survey of Canada. https://en.wikipedia.org/wiki/Geological_Survey_of_Canada, accessed 10/9/2017.

11

CAPITALISM AND NEOLIBERAL GOVERNMENTALITY IN FOUCAULT

Introduction

In this chapter we add an economic aspect to our understanding of Foucault's concept of governmentality, and we explore its widening and deepening in neoliberal governmentality. In other words, one of the ideas in our tool-box is strengthened, and another, neoliberal governmentality, added.

Foucault analyzes both capitalism and neoliberalism historically and specifically through central thinkers of the time. He analyzes capitalism in France in 1754 through 1764, based on the economic theory of the physiocrats,[1] and analyzed in a text by the French economist Louis Paul Abeille. He also analyzes the liberal political economists in Great Britain, Smith and Ferguson. He analyzes the birth of neoliberalism in

Germany and the United States, but I will focus here on the United States, which Foucault approaches primarily through Becker (Foucault 2008:219).

What I hope to clarify in this chapter is the close relationship in Foucault's thought between the economy and government in capitalism and neoliberalism, the similarities and differences between capitalism and neoliberalism, and American neoliberalism's understanding of the subject.

This is Foucault's (2007:341) own summary of the year of lectures of 1977 to 1978: "Basically I have done nothing else for several months but try to provide you with a commentary on these texts on grains and scarcity, which through some detours, was always the issue." This is also the year of lectures that produced the idea of governmentality, which first appeared in the fourth lecture. Governmentality studies, in other words, grew out of Foucault's study of early economists' analyses of grain and scarcity. Foucault argues that the transformation to capitalism is from disciplinary mechanisms characteristic of French governing during the mercantilist era, to "apparatuses of security" characteristic of French governing during capitalism—which he will call governmentality in the fourth lecture—all in response to grain scarcity.

In the following year of lectures, 1978 to 1979, Foucault turns to liberalism and neoliberalism. Liberalism, analyzed through Adam Smith and Adam Ferguson, is associated with governmentality. To be more precise, their political economy is the science that informs governmentality, and in fact limits it to the non-economic. Foucault (2008:286) sees political economy as lateral to government, not a governmental rationality itself. American neoliberalism is also exterior to government; in fact, it constitutes a permanent critique of governmental action (Foucault 2008:246). The transformation is from the science of government to the critique of the government. In both cases governmental action and economic theory are intimately related.

As I argued in Chapter 1, we need to learn to think of power and the economy together. Conceptually we tend to separate all things political from all things economic. But government, any form of government, is always informed by economic theories and critiques. We need to turn to historical and ethnographic details to see how government and economics are exercised together.

Conservation interventions in the developing world have been shaped by development: they share funders, models, methods, etc. And development essentially means economic growth. Development assumes that getting a nation or a community to "develop" is getting them to engage with "the market" so that economic growth will occur. Sustainable development holds that economic growth will halt environmental degradation too. This means that economic thinking underlies conservation. In fact, it underlies conservation increasingly, as conservationists look for ways to translate species and forests and parks into economic value, believing that this will strengthen conservation. Economic theory now underlies much conservation. It has become natural to us to think in these terms. I believe that we need to think differently.

The transformation to capitalism in Foucault

In the lectures of 1977–8, the inseparability of politics and economics in Foucault's thought is particularly clear. The problem of grain scarcity in France, as Foucault presents it, was seen as both economic and political before the physiocrats' edicts, and the transformation afterwards is also both economic and political. The basic transformation is one from a disciplinary state regulating commerce to governmentality, which lifts those regulations because it is informed by the idea of a free, natural economy. This is discipline to governmentality, highlighting the role of the economic. In fact, the whole year of lectures presents this transformation through the economics of grain scarcity. Neither discipline nor governmentality is merely political in Foucault's thought, because politics and economics are inseparable.

Before the physiocrats: discipline

The problem of grain scarcity in mid-eighteenth-century France, as Foucault presents it, was inherently both economic—grain prices were seen as out of control—and political—people were seen as out of control, ready to revolt in the cities because grain prices were out of control. State attempts to control the problem were characteristically disciplinary: a host of legal regulations to prevent food shortages, including price controls, prohibitions on hoarding, limits on exports, prescriptions about amount of land cultivated, etc. The police have only one mode of intervention during this time, and that is to enforce these regulations on grain (Foucault 2007:31–2, 340, 347). This is the police state of the seventeenth and first half of the eighteenth centuries, which aimed to control people in town and the circulation of goods in the market, in relation to each other (Foucault 2007:335). The world economic system at this time is mercantilism, to which Foucault says police are essential. Mercantilism uses commerce as an instrument in intra-European competition, and depends on a large, working population with low wages to draw gold into the state. Foucault characterizes this period as having a number of fundamental elements linked together: *raison d'état* (the state itself has replaced the sovereign as the main reason for the state), urban privilege (the state is concerned primarily with the urban, not the rural, population), and the fundamental link between the police and the primacy of the commodity (Foucault 2007:338). Commerce, town, regulation, and discipline are the most characteristic elements of police practice (Foucault 2007:341).

This disciplinary, regulatory, police-enforced anti-scarcity policy fails, because lowering the price of grain to lower its cost for townspeople also lowers peasant profit, making them sow less land, which makes them more vulnerable to climate risk and actually increases the risk of shortages of grain (Foucault 2007:33).

After the physiocrats: governmentality

The problem of grain scarcity becomes the basis for criticism of the police state by the physiocrats, and directs the transformation they craft (Foucault 2007:347). A series of edicts dated 1754 to 1764, based on their economic theory and analyzed by Foucault in a text by Abeille, have four new theses. First: the physiocrats argue that grain must be well paid for rather than cheap. This thesis shows a set of significant shifts in the essential object of government, with rural agriculture replacing the town, production replacing circulation, and return to primary producer replacing cost to consumer. Second: the physiocrats argue that if grain is well paid for its price will settle at a just price. This thesis calls regulation into question, suggesting that by trying to control things they get worse. It argues that there is a spontaneous regulation of price in the course of things. Third: the physiocrats argue that the size of the population is not a good in itself; like prices, it will adjust itself without regulation (Foucault 2007:345). Fourth: the physiocrats argue that free trade between countries should be allowed, that competition should be between private individuals rather than between countries. In competition between individuals, the good of all will be naturally assured by the behavior of each individual's private interests.

The physiocrats' edicts show a complete change in techniques of government to what Foucault is going to call "governmentality."[2] Under these new governmental techniques, grain, the producers and consumers of grain, and grain in markets are newly seen as having their own free, natural realities. They come to be imagined as "the economy" and "the population." These natural realities are analyzed, and an apparatus is grafted onto, or works within, that analyzed reality, arranging things to limit scarcity without preventing it (Foucault 2007:36–7). Foucault calls this apparatus the apparatus of security, which tries to work within reality to get components of reality to work in relation to each other (Foucault 2007:47). By security, Foucault means a kind of government that manages or facilitates these phenomena newly understood as natural—economic processes or the processes intrinsic to the population—so they do not go off course. Governmentality ensures the "security of the natural phenomena of economic processes or processes intrinsic to population" (Foucault 2007:353). Foucault uses the economic term "laissez-faire" to describe the governmental state's new role: "It will be necessary to arouse, to facilitate, and to *laisser faire*, in other words to manage and no longer to control through rules and regulations" (Foucault 2007:353). This physiocratic policy works: prices do rise with abundance, as long as people are allowed to hoard and to export (Foucault 2007:37). In terms of grain, the transformation is one from preventing grain scarcity through regulation and police enforcement, to securing the free circulation of grain through the new science of economics.[3]

Foucault writes that the founding act of economic thought and the fundamental principle of economic government is the free circulation of grain (Foucault 2007:34). He is arguing that economics as a discipline and the economy as a concept grew from the physiocrats' thinking about grain circulation. This "moment" is profoundly linked to capitalism: "This ideology of freedom really was one of the conditions of development of … capitalist forms of the economy" (Foucault 2007:48). Knowledge becomes indispensable to government after the physiocrats: a quite peculiar relationship of power and knowledge, government and science, appears (Foucault 2007:351). This science of government, first called political economy, becomes economics.

The free circulation of grain as central to governmentality also allows a spatial distinction with discipline. Discipline "isolates a space," it "concentrates, focuses, and encloses." "The first action of discipline is in fact to circumscribe a space in which its power and the mechanisms of its power will function fully and without limit." (Foucault 2007:44–5). It is protectionist, and this protectionism is focused on the space of the market. In contrast, the apparatuses of security (governmentality) "have the constant tendency to expand; they are centrifugal." New elements are continually being added (even more so with American neoliberalism); governmentality allows "the development of ever-wider circuits" (Foucault 2007:45).

Foucault (2007:49) links the free circulation of grain tightly to the new technology of power, a technology in which power is thought of as regulation that can only be carried out through freedom. In the eighteenth century, freedom is:

> no longer the exemptions and privileges attached to a person, but the possibility of movement, change of place, and processes of circulation of both people and things. I think it is this freedom of circulation, in the broad sense of the term, it is in terms of this option of circulation, that we should understand the word freedom.
>
> *(Foucault 2007:48–9)*

Freedom is both an ideology and a technique of government (Foucault 2007:48), and it is "the correlative," "one of the facets, aspects, or dimensions" of the deployment of apparatuses of security, regulation "through and by reliance on the freedom of each" (Foucault 2007:48, 49). Freedom is a technology of power (Foucault 2007:49).

Freedom as a technology of power transforms police into what I argue is policy. In discipline, the police are essential to enforce the myriad regulations of the market. With governmentality, Foucault (2007:353–4) says that there must be a double system. Many of the previous functions of the police fall out of their domain, most notably the regulation of the economy and the population. The police are left with the "simply negative function" of preventing disorder. All of their old positive functions are taken up by "a whole series of institutions, apparatuses, mechanisms, and so on," "the great mechanisms of incentive-regulation: the economy, management of population, etcetera" (Foucault 2007:354). Economic practice and population management, together with a law constructed to respect freedom, eclipse disciplinary regulations and police. I think economic practice and population management become what we now call policy.

Liberalism and governmentality

The physiocrats and the circulation of grain, ushering in the transformation from discipline to governmentality, occupy Foucault in his 1977–8 lectures. In the following year, 1978–9, he is talking about the same transformation using the term liberalism, and basing his lectures on the Englishmen Adam Smith and Adam Ferguson. This allows him to contrast liberalism with German and especially American neoliberalism, based on the work of the radical neoliberal Gary Becker in the 1970s. This introduces another, distinct transformation. The 1978–9 lectures are also the closest Foucault gets to contemporary politics.

The key elements of the physiocrats' critique of government are present in Foucault's discussion of liberalism, and other elements are added or clarified. The government must be laissez-faire with the market; in short, the government must not govern too much (Foucault 2008:319–20). Foucault (2008:278–82) analyzes Adam Smith's idea of the "invisible hand," stressing the invisibility rather than the hand to argue that the economists saw the economy as opaque and non-totalizable. Thus, the government must not govern too much because it cannot know the economy. Adam Smith's economic liberalism in fact disqualifies the government from any efforts to govern the economy. The government instead governs civil society, also called simply "society" or "the nation," which is also what Foucault in 1978 called the population (Foucault 2008:295–6).

In liberal economics, the government must be laissez-faire with *homo economicus* too. In Adam Smith, *homo economicus* "is someone who pursues his own interest, and whose interest is such that it converges spontaneously with the interest of others." This is the invisible hand (Foucault 2008:270, 278). It makes *homo economicus* "the partner, the vis-à-vis, and the basic element of the new governmental reason formulated in the eighteenth century [governmentality]" (Foucault 2008:271). In Ferguson, according to Foucault (2008:298), civil society is "the concrete, encompassing element within which the economic men Smith tried to study operate." In liberal governmentality, *homo economicus* and civil society "belong to the same ensemble" (Foucault 2008:296).

American neoliberal governmentality

The key to the transformation from liberalism (or governmentality as defined in 1978) to American neoliberalism (or neoliberal governmentality) is that *homo economicus* is transformed from a subject that government must let be to a subject who is eminently governable (Foucault 2008:252; Gordon 1991:43). The essential nature of the economic subject, or at least of his (or her) relation to government, is thus reversed. Another important difference is that, although liberal economics disqualifies the government from governing the economy, American neoliberalism goes further, constituting a "permanent criticism" of government: it will not let government be, scrutinizing all of its actions in economic terms (Foucault 2008:246–7).

A third important characteristic of American neoliberalism is its desire to extend the rationality of economic analysis to formerly non-economic domains, including labor. This differentiates it from German neoliberalism, which sees the regulation of prices by the market as the only basis for a rational economy, but also as such a fragile process that it must be supported "by an internal and vigilant policy of social interventions (involving assistance to the unemployed, health care cover, a housing policy, etcetera)" (Foucault 2008:323). So American neoliberalism extends its rationality outwards to non-economic realms, and German neoliberalism focuses inwards on social interventions to support the market regulation of prices.

Freedom is at the heart of both liberalism and neoliberalism. The physiocrats argued for the free circulation of grain and free competition between individuals. This freedom, as Foucault argues, is both essential to understanding capitalism and a technology of power. Neoliberalism strengthens the ideology of freedom, and also deepens its governing of individuals.

Foucault uses the American neoliberal view of *homo economicus*, the economic subject, to demonstrate how neoliberalism thinks about governing the individual. The economic subject was seen as a "person whose conduct is influenced by the gains and losses associated with it" (Foucault 2008:259). This means that attempts, for example, to prevent crime, must act on the "market milieu" in which the subject performs their criminal conduct such that their losses outweigh their gains. "There is an environmental type of intervention instead of the internal subjugation of individuals" (Foucault 2008:259–60). Government "will act on the environment and systematically modify its variables," because the economic subject "responds systematically to systemic modifications artificially introduced into the environment" (Foucault 2008:271). This is precisely what the integrated conservation and development projects in Chapter 10 were hoping to do: provide agricultural inputs to increase profit thus protecting the bordering park, where they also increased guards, from farmers. The projects acted on the local milieu to increase the benefits of farming outside the park and increase the costs of farming inside the park.

The physiocrats reacted against government regulation of grain; American neoliberals, in Foucault's view, reacted against three elements: Keynesian policy, the social pacts of war, and the growth of the federal administration through economic and social programs (Foucault 2008:217). In the first element, Franklin D. Roosevelt's New Deal Keynesian policy was crafted in 1933–4, 1932 having been the worst year of the depression in the US. The policy was criticized in 1934 by Henry Calvert Simons, who Foucault sees as the father of the Chicago school of economics. The second element refers to the British Beveridge plan of 1942, which basically laid out the welfare state of social and economic interventionism as reward for people's sacrifices. This plan was influential in the US. It was also criticized by Simons. The third element is the history of US state interventionism, from Truman through Johnson, which involved growth in the size of the state (Foucault 2008:216–17). It wasn't interventionism that the neoliberals objected to, but the wrong kinds of interventions.

Government works at two levels, the population and the subject. Though the competing individual is present in the liberal theses, he (or she) is to be left alone, and the population, civil society, is primary. By the 1950s, American neoliberals led a shift in point of view to individual subjects through the idea of human capital, which emerges from their efforts to understand labor (Foucault 2008:252).

According to Foucault, neoliberals were "trying to introduce labor into the field of economic analysis" (Foucault 2008:220). Neoliberals argued that economic theory left labor unexplored or reduced to time (Foucault 2008:219–20). Foucault (2008:220) contrasts their approach to Marx, while noting that "neo-liberals practically never argue with Marx." Marx, he says, shows that capitalism reduces labor to labor power and time; neoliberals argue that it is not capitalism itself but the economic theory of capitalism that abstracts and reduces labor (Foucault 2008:221). Early economic theory studied the mechanisms of production, exchange, and consumption; neoliberals, in contrast, studied the ways in which scarce means were allocated to competing ends (Foucault 2008:222). Allocation was carried out by individuals; thus, "the starting point and general frame of reference for economic analysis should be the way in which individuals allocate these scarce means to alternative ends" (Foucault 2008:222). The neoliberal analysis of labor, then, concerns how a working person uses the means available to him or her. The individual becomes "an active economic subject" (Foucault 2008:223). The neoliberals' economic subject is an entrepreneur, an entrepreneur of him or herself (Foucault 2008:226). And rational government should be modeled on the economic rationality of the governed (Foucault 2008:212–13).

From the workers' point of view, wage is income, the income of a capital that is the worker's ability or skill. So, neoliberals argue, income comprises a capital, though it is inseparable from the person—thus "human capital." Human capital also marks the extension of economic analysis into a previously non-economic domain (Foucault 2008:219). Neoliberals see innovation, which keeps the rate of profit from falling, as a product of human capital (Foucault 2008:231). Investing in human capital, then, accounts for the growth of Western economies (Foucault 2008:232). The idea of improving human capital leads early neoliberals to eugenics (innate human capital), but they later turn to acquired human capital, and, thus, education, health, and migration (Foucault 2008:228–30). In other words, the sorts of governmental interventions that neoliberals supported, in education, health, and migration, were intended to increase human capital, and, thus, innovation, and profit, in America. In sum, American neoliberalism focuses on subject creation as investment in human capital.

Conclusion

In Foucault's thought, as I hope I have demonstrated, government and economy are very closely related. This is true during the mercantilist, disciplinary period, when the state regulates commodities; during the capitalist, governmental and liberal period, when the state shrinks and backs off from the economy informed by the new knowledge of economics; and during the neoliberal period, when neoliberals become the constant critics of governmental practices. I think we need to learn to think of government and economics together.

One of the primary differences between the liberal state and the neoliberal state is that the former focuses on the civil society or population while the latter focuses on the individual subject. Governmentality governs the population (for its own welfare) largely by securing the freedom of the economy. Neoliberal governmentality governs subjects in two ways, by improving their acquired human capital, thus ensuring the innovations that allow for economic growth, and by manipulating the milieu in which economic subjects calculate the costs and benefits of their conduct.

Conservation occurs in a world defined by neoliberal governmentality. It seems natural to conservation policy-makers and implementers to see the people with whom they cross paths as economic subjects, albeit with low human capital. It is easy for us to label hunters as poachers, for example, without examining the question of whether they are hunting to eat or hunting for sale (which could also, of course, mean hunting to sell in order to buy something to eat). It is easy for us to believe that we can address this by introducing alternative sources of income and stepping up the costs of poaching. We act on the economic milieu of people we imagine as economic subjects in order to govern their conduct. We need to become aware of what we are doing. Questions must emerge: are these people really reducible to economic subjects? As free subjects, they are also free to think in their own ways: are their calculations of costs and benefits what we expect? Are they even making these sorts of calculations? Tania Li (see Chapter 10) would argue that the hill people in Sulawesi are making calculations, and they know that the benefits we have to offer don't outweigh what they're giving up in the park. Anna Tsing (see Chapter 7) would argue that the Meratus Dayak of Kalimantan know how to live in the forest, so we need to sit down and learn from them. I think we need to be capable of opening our minds to these possibilities, and to other ideas outside the grids of neoliberal economics.

Notes

1. Physiocracy is an economic theory developed by a group of eighteenth-century Enlightenment French economists who believed that the wealth of nations was derived solely from the value of agricultural production. It is one of the first well-developed theories of economics. It immediately preceded the first modern school, classical economics, which began with the publication of Adam Smith's *The Wealth of Nations* in 1776. Mercantilism, which preceded physiocracy, focused on the ruler's wealth, accumulation of gold, or the balance of trade (Wikipedia).
2. In the second lecture of 1978, Foucault names these new techniques "apparatuses of security"; by the fourth lecture these apparatuses serve a sort of government he newly terms governmentality. By the last lecture of that year, governmentality means "governmental reason," and there are multiple governmentalities, including disciplinary governmentality and new or modern governmentality (Foucault 2007:354). I will retain the meaning of governmentality used in my Chapter 5.
3. Scarcity in general, as a massive force, comes to be seen as a chimera induced by regulations to prevent it, but scarcity that "causes the death of individuals not only does not disappear, it must not disappear" (Foucault 2007:41–2). There is a gap between the new free and natural population and the people who are not relevant (Foucault 2007:44).

References

Foucault, Michel 2007 *Michel Foucault: Security, Territory, Population: Lectures at the College de France, 1977–1978.* Michel Senellart, ed. Graham Burchell, trans. New York: Picador.

Foucault, Michel 2008 *The Birth of Biopolitics: Lectures at the College de France, 1978–1979.* Michel Senellart, ed. Graham Burchell, trans. New York: Picador.

Gordon, Colin 1991 Governmental Rationality: An Introduction. In *The Foucault Effect: Studies in Governmentality with Two Lectures by and an Interview with Michel Foucault.* Graham Burchell, Colin Gordon, and Peter Miller, eds. Pp. 1–51. Chicago: The University of Chicago Press.

Wikipedia, The Free Encyclopedia n.d. Physiocracy. https://en.wikipedia.org/wiki/Physiocracy. Accessed 12/5/2019.

12

CULTIVATING NEOLIBERAL SUBJECTS IN ETHNOGRAPHIES

Introduction

Neoliberalism is often defined as free trade, deregulation of markets, privatization, no state welfare provision, and individuation (Wikipedia n.d.). According to Foucault (as we have seen in Chapter 11), individuation is *homo economicus*, the idea of a person whose conduct can be shaped because it is influenced by the gains and losses associated with that conduct (Foucault 2008:259). This means that government must act on the "market milieu" in which the subject performs a conduct, systematically modifying its variables, in order to shape that conduct (Foucault 2008:271). Individuation meant that groups and communities were ignored by governments and other organizations influenced by neoliberalism. But then, and remembering that Foucault highlighted its expansionist tendencies, neoliberalism expanded into community.

Tania Li (2007:232–4), basing her idea on Nikolas Rose (1999), explores the connection between neoliberalism and "government through community." What Rose means, Li explains, is that issues are problematized and solutions proposed in terms of community features and dynamics. These features include emotional relationships, individual identities, and cultures of values and meanings. Note that culture is part of this imagination of community. Paradoxically, community is both assumed to be natural, and needs to be improved, giving Li's improving trustees work to do that is believed to be enabling or facilitating rather than dominating. In the developing world, Li (2007:234) says:

> In the distinctly neoliberal formulation of the World Bank, communities of poor people were encouraged to take on responsibility for their own improvement by engaging with markets, learning how to conduct themselves in competitive arenas, and making appropriate choices.

In practice, communities were rendered technical through participatory rural appraisal techniques, which directed them to solve their own problems, which policy-makers and implementers imagined as local.

The formalization or creation of user groups is a related technology of participatory development and conservation. Li (2007:235) describes the popularity of user groups:

> To construct an arena of intervention [to render participation technical], experts had to identify or create groups.... In this spirit, groups were made visible, formalized, and improved where they already existed, [and] crafted where they were absent.... In 2001, development experts enthusiastically announced [the attainment of] "some 408,000 to 478,000 groups emerging with 8.2–14.3 million members in watershed, irrigation, microfinance, forest, and integrated pest management, and for farmers' research."

Examples include the forest councils analyzed by Agrawal (see Chapter 10).

The idea of social capital began to emerge in development in the same time period. This began with microcredit, which originated in 1983 with the Grameen Bank in Bangladesh.[1] It combined the user group idea, here usually a group of borrowing women, with the philosophy that the groups' "social capital" could replace assets as loan collateral. The small loans were intended to shape the women into entrepreneurial subjects, who would then develop themselves and their families. We will examine a microcredit project in Bolivia below. In community-driven development (CDD), social capital was a way for social scientists to convince World Bank economists to consider a community-based project in Indonesia, which we will consider in detail below. This time, the idea was that social capital could be improved by arranging for communities to compete for grants. Communities were treated as subjects to be improved by economic means, especially competition. Once improved, communities would be capable of developing themselves. Essentially, social capital allowed economists to view communities in the way that human capital allowed them to view individuals. We might argue that the concept of social capital created a sort of social *economicus*.

James Ferguson has described neoliberalism as a "new governmental rationality" that seeks not only to reform states but to create "new modes of subjectification," specifically new citizen and community subjects that are responsible for government results without direct state intervention.[2] Microcredit and CDD represent examples of these new modes of subject formation.

The need for microcredit and CDD—which Ferguson suggests were linked to neoliberal subject formation—was increased by another neoliberal project: the Washington Consensus, also known as structural adjustment. The Washington Consensus is an International Monetary Fund (IMF)/World Bank policy that requires developing countries who want a loan from them to implement some or all of the following policies: cutting government spending on things like welfare programs to reduce the budget deficit; raising tax revenues and trying to improve tax collection by clamping down on tax avoidance; control of inflation, usually through higher interest rates; privatization of state-owned industries; de-regulation of markets to encourage competition; opening the economy to free trade (removing tariff barriers which protect domestic industries); ending food subsidies; and devaluation of currencies to restore competitiveness. Structural adjustment exacerbated poverty as states were forced to move away from policies that aided the poor and towards policies like privatization that passed costs on to the poor. Demand for projects that were seen as creating subjects who could help themselves increased (see Lazar 2004; Molyneux 2002).

Microcredit and CDD are interventions designed to create newly economic subjects, but there are other, less direct ways in which the market milieu could be shaped to cultivate new subjectivities. The contracting, sub-contracting, and sub-sub-contracting supply chains so common in the contemporary global economy have subject formation characteristics themselves, as Anna Tsing argues. I will explore the new subjects she describes below. Her ideas demonstrate that neoliberalism has expanded into culture as well as society. Government through community, as argued by Li through Rose, also depends on community features of emotional relationships, individual identities, and cultures of values and meanings. In Tsing's ethnography, firms in supply chains make the same sorts of features do the work of disciplining individual workers.

User groups have been associated with conservation for a long time, particularly in forest user groups like the forest councils described by Arun Agrawal (see Chapter 10). CDD can have any goal attached to it, and conservation is one of those goals (e.g., Russo et al. 2004). Any conservation project that is integrated with development may use microcredit. The larger point is that most interventions today that act on people have some component that is seeking to create neoliberal subjects. We do not know if they are succeeding in this goal, and we do not know if neoliberal subjects are more or less consuming of the environment. In addition, Tsing's piece raises the probability that the market itself is changing subjectivities.

Not all neoliberal interventions intend to create individual or social *homo economicus* who will freely compete in the market. As Tania Li (2014) argues, neoliberal ideas about people are combined in assemblages (see Chapter 16) with earlier, liberal ideas about people—particularly in conservation assemblages—such that some indigenous and highland farming populations are actually protected from some market processes. In these cases, as in *The Will to Improve* (2007), the subjects are communities rather than individuals. This is particularly relevant to conservation interventions.

Subjects in microcredit

Microcredit "is a form of development intervention perfectly in line with neoliberal philosophies of the entrepreneurial, individual citizen" (Lazar 2004:302). Sian

Lazar[3] (2004:302), based on ethnographic research, argues that three microcredit NGOs working with Aymara migrants in El Alto, Bolivia, are undertaking a governmentality project: seeking to create "'empowered' individual, entrepreneurial, active citizens who will take responsibility for their own and their families' welfare, and who are prepared for the market rather than the state to provide for them." All three are based on the Grameen Bank, which first popularized the idea of social capital. The kind of governmentality project the NGOs are undertaking is clearly subject formation. Lazar (2004:301) also sees the three NGOs as operating from the basis of very neoliberal ideas of citizenship in which individuals act as economic subjects, essentially *homo economicus.* So the microcredit NGOs are neoliberal governmentality projects attempting to create neoliberal subjects.

Except in the narrow terms of loan repayment, the microcredit project she describes is a failure. Most loans are misused and repaid through a previously existing informal loan system (which has higher interest rates but much more flexible deadlines), trainings are mostly ignored, and meetings are used for women's own purposes. Nothing suggests that the women have become more entrepreneurial than they already are (Lazar 2004:306).

In fact, the narrow success of loan repayment is not the success it appears to be: microcredit groups "rely upon the women's existing networks of family and friends, and associated cultural understandings and obligations, in order to ensure loan repayment" (Lazar 2004:306). As Lazar explains, the NGOs' neoliberal rhetoric emphasizes a rationality that disembeds the economic from the social.[4] In practice, however, they "navigate effectively between disembedded and embedded economic rationalities," as do the Aymara women (Lazar 2004:306). Their success with loan repayment stems from "their contradictory reliance upon the embedded economic rationalities that, on the surface, they seek to modify" (Lazar 2004:316). They're depending on the social to guarantee the economic, but they say they're creating market-savvy entrepreneurs. In a very real sense the NGOs are relying on the Aymara women's social capital for project success.

The subject in community-driven development

CDD is a World Bank policy model, also used by other organizations, with hundreds of active projects all over the developing world (The World Bank Group 2004). It is a scaling up of a huge and still on-going World Bank project in Indonesia called KDP (Kecamatan, or community, Development Program). KDP was designed by the anthropologist Scott Guggenheim,[5] and began just before the end of the New Order government in 1998. For Guggenheim, social capital was simply a translation of the "community" that anthropologists study into economic terms understood by the World Bank. For World Bank economists, in other words, the idea of social capital made sense out of an anthropologist's ideas about improving communities.

Guggenheim did not design KDP to make communities into neoliberal subjects. But this is what Tania Li,[6] based on ethnographic fieldwork, believes the project is doing. The KDP offered "a carefully designed process for planning and managing projects, and placed that process in the hands of villagers" (Li 2007:231). Li argues that "the village level planning process devised by the bank, backed by its rule books, monitors, and auditing devices, was designed to *shape* desires and act on actions, setting the conditions so that people would behave as they ought" (Li 2007:231, author's italics). This is a nice definition of governmentality. It is also neoliberal: "In the distinctly neoliberal formulation of the World Bank, communities of poor people were encouraged to take on responsibility for their own improvement by engaging with markets, learning how to conduct themselves in competitive arenas, and making appropriate choices" (Li 2007:234). Like microcredit but at the level of the community, CDD is neoliberal governmentality at work.

KDP provided grants of $60,000–$110,000 to subdistricts, where a committee (including villagers, facilitators, and officials) judged competing projects for infrastructure or small enterprise credit that were proposed by groups of villagers. Competition for funds was the cornerstone of the project. In order to enter the competition, villagers had to follow rules about unfamiliar things like forming committees, holding public meetings, and monitoring projects, which were detailed in manuals and presented verbally by over 4,000 consultants and facilitators working at the village and subdistrict level. In addition, a man and woman from each village were trained and given stipends. There was a menu of options that villagers were encouraged to choose for their competing projects. (Li 2007:247–9) Project success was measured by the "uptake" of KDP practices and ideas, rather than the infrastructure built or the funds dispersed (Li 2007:253). KDP practices and ideas were essentially neoliberal: "transparency, accountability, and efficiency" (Li 2007:247).

The goal of KDP was to improve the social capital of communities in order to "supply village infrastructure more efficiently, alleviate poverty, promote economic growth, foster good governance, and enhance local capacities for conflict management" (Li 2007:244). It also has a utopian aspect: improved communities, it argued—and the project was designed before the New Order government fell—could spread good governance up to the national level, reforming the state (Li 2007:1–2). The assumption is that improving communities can not only lead people to continue to improve themselves, but can also lead to the improvement of national-level problems. In KDP, "community" is simultaneously something that is natural and the target of intervention: neoliberal practices and ideas would build on an authentically local base. Here the economic is assumed to "improve" the social, which makes sense because it is seen in fundamentally economic terms, as social capital. This is more ambitious than microcredit: microcredit builds on existing social capital, whereas KDP intends to increase social capital.

Li's deepest criticism of KDP is that it ignores structural economic inequalities. The structural inequalities that plague many Indonesian villages stem from the extraction of natural resources, particularly forests (but also mining, and forest destruction for palm oil plantations), by a state that claims to own forests and benefits from their extraction or destruction—and is aided in that pursuit by the World Bank, which continues to support ventures in forestry, mining, and plantation agriculture (Li 2007:263). In some places, like the uplands of Sulawesi, conservation ironically multiplied the effects of long-term state extraction of natural resources. This has led to violent conflicts over natural resources. KDP was applied in these contexts too (in a version called SPADA–Support for Poor and Disadvantaged Areas) to replace competition over scarce resources with competition over funds (Li 2007:256–62). Neither the conservationists anxious to conserve any land left untouched by timber or palm oil companies, nor the World Bank designers of SPADA, felt that they could address causal structural inequalities. Li cites from a SPADA document: "The bank is not in a position to influence directly the two immediate causes of conflict: organizational and resource grabs by national and regional elites, and the active or passive role of the armed forces in promoting and resolving conflict" (Li 2007:262). The fact that the World Bank has not halted its own support for forestry, mining, and plantation agriculture businesses means, as Li (2007:267) asserts, that it cannot see neoliberalism as a cause of poverty and conflict, only as a solution.

If Li is right, and I do agree with her, then neoliberalism is posing itself as the solution—through community and society—to large-scale problems caused or at least supported by neoliberalism—problems that are not community or society based. Li also argues that KDP undermines the ability of its subject communities to demand change. Neoliberal empowerment empowers communities to demand better infrastructure or better governance; it does not empower them to demand better access to land, or fair prices or wages (Li 2007:256). It makes them into "good villagers" (Li 2007:254). She ends her discussion of KDP by quoting an Indonesian critic who says that "democracy, public participation, accountability and social and economic rights are all historically tied to the outcome of struggles of social forces and interests" (Li 2007:269).

To take a step back, Li has used a Foucauldian analysis of governmental power and neoliberalism in order to understand how a project defined by that sort of power is preventing a Marxist sort of revolution. I agree with her. And I believe it is relevant to conservation. Conservation should have stepped in earlier to bring pressure to bear on state extraction of natural resources in Indonesia (and elsewhere), rather than claiming the land that was left for conservation, which multiplied the effects of state extraction on the local population.

Non-market neoliberal subjects

Conservation interventions are particularly linked to the creation of communities as neoliberal subjects who are counter-intuitively also outside the market. As Li (2014:34–5) says, based on her field research, *homo economicus* minus the market is "the curious, collective subject that figures in contemporary land regimes and conservation initiatives." These interventions particularly target "indigenous" or "tribal" populations and highland farmers (Li 2014:42). This non-market yet neoliberal subject emerges from the combination of older, liberal ideas about colonial subjects, especially those seen as tribes (versus peasants, farmers, or the urban educated), with neoliberalism.

Liberal colonial regimes found tribal subjects to lack the capacity to survive the market, and thus sought to protect them from it. Colonial authorities thought "they should be governed in terms of their difference: fixed in their alterity, fixed in place on ancestral/customary land, and fixed in the sense of repaired, with inappropriate changes reversed or abated to restore them to their authentic selves" (Li 2014:38). This was a liberal regime, as Li emphasizes: tribes were governed "in accordance with the grain of things," in accordance with research about tribal population (research that was, however, "not ethnographic in the modern sense") (Li 2014:40, 42). Expert knowledge found tribal subjects to be collective. And the law was often used as a tactic; for example, land sales were forbidden. This left populations identified as tribal without individual land tenure, setting the stage for future natural resource extraction by the state.

Much more recently, and rather ironically, these paternalistic policies were "appropriated and made into demands" by "indigenous" people (see Chapter 8) in order to secure land tenure (Li 2014:40). But they were also incorporated into assemblages dominated by neoliberalism, in which the persistence of non-market fixes is surprising and awkward (Li 2014:42–3). Neoliberalism still treated these populations as "communities," but with a "neoliberal twist: they are communities capable of exercising their 'free, prior, informed, consent'" (Li 2014:43). This idea exists alongside the assumption that "indigenous people have desires, habits and beliefs that favour collectivity, equity, environmental sustainability, and conservation," meaning that "they will reject practices that damage land, pollute water or destroy forests. They won't privatize their collective land or plant it with lucrative cash crops, take on debt, or sell land to the highest bidder" (Li 2014:44).

Conservation linked to global warming and REDD+ (referring to Reducing Emissions from Deforestation and forest Degradation in developing countries) intensified these interventions. REDD+ programs like these populations, because they are believed to be naturally aligned with project goals. They are, however, uncertain about free, prior, informed consent: they give communities choices, but only among REDD+ alternatives. As Li (2014:46) explains:

> Only "forest people" and indigenous people who have been collectivized, arborealized, fixed in place, and fixed in their difference could be encouraged, empowered and potentially coerced to make a non-market choice—one that commits them to making less money than they could with other "options."

It seems to me that REDD+ reveals an essential contradiction between conservation and the neoliberal subject: the conservation choice is the non-market choice.

Subjects in supply chain capitalism

Microcredit and CDD are interventions designed to create neoliberal subjects, but there are other, less direct ways in which the market milieu may be shaped to cultivate new subjectivities. The supply chains so common in the contemporary global economy have subject formation characteristics themselves, as Anna Tsing[7] argues. Tsing (2009:148) coins the term "supply chain capitalism" for "commodity chains based on sub-contracting, outsourcing, and allied arrangements in which the autonomy of component enterprises is legally established even as the enterprises are disciplined within the chain as a whole." Supply chain capitalism is an attempt to describe the modern, global economy, like David Harvey's (1990) flexible accumulation, which also emphasizes contracting. Tsing's (2009:150, 158) point is that supply chains depend on subject formation based on non-economic factors: gender, ethnicity, nationality, religion, and citizenship status; in short, social and cultural differences.

I argue that there is a great deal about supply chain capitalism that is neoliberal, especially attention to subject formation, governing indirectly through a milieu (the supply chain), and expansion into the non-economic, but there is also perhaps a critical difference between neoliberalism and this sort of capitalism.

Diversity structures capitalism by linking diverse niches together to create global chains, and, of most interest here, "diversity conditions the responses of both capital and labor to the problems of cutting labor costs and disciplining the workforce" (Tsing 2009:150). Much of this diversity emerges from outside the chains themselves: "No firm has to personally invent patriarchy, colonialism, war, racism, or imprisonment, yet each of these is privileged in supply chain labor mobilization" (Tsing 2009:151). In addition, "in a time of neoliberal globalization," Tsing (2009:150) explains, "[supply chains] are often formed in legal gray zones and within the constant flux of boom-and-bust opportunities." The diversity of chains and the diverse "figures" performed in chains, in other words, cannot be governed from inside the chain, which makes it interesting. Walmart, for example, created the figure of the servant-leader, but other figures "enter the supply chain obliquely, used but not created by capital" (Tsing 2009:157).

Supply chains succeed in avoiding high labor costs by outsourcing labor and distancing it outside historical labor struggles. This makes non-work representations of labor—in which work appears, for example, as management, consumption, or entrepreneurship—"key features in defining supply chain labor" (Tsing 2009:151). Workers perform these "non-work tropes" or "figures of labor." These new tropes or figures of labor can be seen as new subjectivities that workers are encouraged to embrace or old subjectivities that workers are encouraged to see as work. Labor is both recruited and disciplined by these figures that are outside the economic, outside work, and often outside the chain (Tsing 2009:157, 158).

Tsing's first example of a "figure" is the Manchester industrial worker whose struggles inspired Marx and Engels and led to a politics based on the working class (Tsing 2009:153). This is essentially the figure that supply chain capitalism avoids with non-work figures. She then briefly introduces three corporate images: General Motors and Fordism, McDonald's franchise arrangements, and Walmart (Tsing 2009:154–6). These three figures "have shaped the making of the global scale by inspiring boosters, participants, and critics to imagine the bigness of capitalism" (Tsing 2009:154). The Fordist dream of standardized production separated a forward-looking economy from a less and less relevant culture (Tsing 2009:155). McDonald's represented the bigness of capitalism as local expressions of global standards, and allowed people to imagine the "global homogenization of economic concepts, rules and procedures" (Tsing 2009:155).[8] Walmart is the first example of a supply chain capitalism figure of labor. Its success, Tsing posits, has to do with a clear demarcation between "what Wal-Mart wants to control (e.g., prices, marketing, logistics) and what Wal-Mart does not want to control (e.g., labor arrangements, environmental practices, subcontractors' investment strategies)" (Tsing 2009:156). The company standardizes what it wants to control, and contracts out what it doesn't. Labor inside Walmart stores is standardized and controlled; the labor arrangements of contractors are not.

Walmart represents labor in its stores as management or as family: Walmart's "servant-leader" is a (male) store manager who adopts largely female "associates" as "family" (Tsing 2009:159). Female associates are understood to be mothers first and workers second; providing work to women that is part-time, irregular, and poorly paid can then be seen as accommodating their needs as mothers. Coaching and shaming are also used. Tsing argues that Walmart is using patriarchal values: "Gender discrimination *makes labor possible* in the Wal-Mart model" (Tsing 2009:161, author's italics).

The Walmart pattern is even more exploitive in sweatshops, where immigrant entrepreneurs, following the "fiction of contractor independence," recruit other immigrants and "bring performances of ethnic niche specificity into the chain" to discipline and underpay them (Tsing 2009:162). Dreams of consumption, like the ownership of Nike shoes, can supplement dreams of entrepreneurship, like those of FedEx drivers.

One of Tsing's central arguments is that laborers in supply chain capitalism superexploit themselves. She defines superexploitation as "exploitation that depends on so-called non-economic factors" like gender, race, ethnicity, nationality, religion, sexuality, age, and citizenship status. It is also "exploitation greater than might be expected from general economic principles," including, I think, neoliberal economic principles (Tsing 2009:158). Superexploitation is neoliberal *homo economicus* trying to sell himself or herself by enacting cultural and social subjects. But those efforts lower workers' capital, make what they have to sell worth less, rather than raising it, as in Foucault's discussion of neoliberal human capital (see Chapter 11). Performances of labor, Tsing writes, "bring them contracts *and* make it difficult for them to negotiate the wage outside niches for gender, sexuality, and race" (Tsing 2009:159). Though attention to subject formation, governing through a milieu, and expansion into the non-economic are neoliberal through and through, the exploitation piece is either something new within neoliberalism or just plain capitalism.

Tsing's interest in supply chain capitalism is the human condition, as made clear in her essay title, "Supply Chains and the Human Condition." She sees enacting new figures of labor as deeply exploitive. But for Tsing, ever the optimist, "diversity is both the source of low wages, and, potentially, the source of creative alternatives" (Tsing 2009:157). "It both makes supply chain capitalism work and, upon occasion, gets in its way" (Tsing 2009:171). Supply chain capitalism is disorganized, poorly controlled, on the edge of legitimacy, rent by scandals—and these characteristics present openings for criticism and mobilization (Tsing 2009:172). Non-work identities discipline labor but also open alternatives, as cultural diversity tends to do.

Conclusion

Subject creation pervades development, with a long history of user groups, social capital, microcredit, and CDD. It also pervades our economy, especially its low-wage sectors. In both locations, it is neoliberal, based on a core assumption that poor people can be improved by guiding them to become *homo economicus*. But these readings suggest that *homo economicus* is disturbingly cultural and social. Tsing's culture-based labor figures suggest that the non-economic is being folded into the global economy. Li's KDP and the CDD model it inspired suggest that neoliberalism is posing itself as the solution to problems that it has had a hand in creating. Lazar's NGOs suggest a pervasive confusion about the separation between the economic and the social. Neoliberal subject creation, if these authors are right, is using and in fact depends on the non-economic. Yet the people guiding this effort misread the social and the cultural in economic terms. And if Tsing is right, culture is hard to control. And if Lazar is right, society is hard to control too.

Lazar argues that microcredit groups (in spite of the feminist NGO project of empowering women for effective political participation) are apolitical, while other groups women belong to, like the market sellers' union, are political. Loan groups, in other words, may actually pull women from political groups. Li says that KDP makes power (the conduct of senior officials, investors, and the military) and structural economic deformities (capitalist enterprise and the search for profit) invisible, and does not allow change based on these forces to occur. Quoting an Indonesian activist, Li (2007:269) writes that real change is "grinding social change over centuries, colored by often violent and bloody confrontations, not least between social classes." Tsing argues that supply chain capitalism is bad for the "human condition," because it leads workers to superexploit themselves and lowers their human capital. Lazar's politics occur largely inside communities, Tsing's in supply chains, while Li's structural economic inequalities are non-local. KDP communities are expected to improve themselves in spite of the fact that their poverty does not have a local source. What Li clearly wants, and Lazar too I think, is political resistance. So does Tsing, who says that labor figures create openings for criticism and mobilization.

Li (2014) does something quite different with society and culture. Colonial knowledge of certain colonial subjects perceived their society as communal and their culture as incapable of surviving market risks. More recently these same sorts of populations have come to be seen as "indigenous," still communal, forest lovers. Interventions like REDD+ attempt to create communal subjects who will choose what REDD+ has to give, which means making a non-market choice for conservation. Interventions base this on assumptions about their society and culture.

Conservation interventions are working to create neoliberal subjects with and without market sensibilities. Market sensibilities price natural resources; they are responsive to market demands for natural resources. I have a hard time understanding why a conservation intervention would want to create economic subjects. The market itself is already spreading this perspective everywhere (see Chapter 13). Li (2014:46) calls non-market sensibilities, on the other hand, an assumption or a "fragile hope." Non-market but neoliberal subjects are being asked to accept a lot less money through REDD+ than timber is worth. This is the same bargain farmers were asked to accept in Li (2007): accept agricultural inputs rather than farm better land (which they were already farming) in the park. In that case they invaded the park. I believe that economic subjects and conserving subjects do not go hand in hand.

Notes

1. 2005 was the year of microcredit; in 2006 the bank and its founder (Muhammad Yunus) were awarded the Nobel Peace Prize.
2. I am paraphrasing James Ferguson (2010:172), who cites Barry, Osborne, and Rose (1996) and others.
3. Sian Lazar is a senior lecturer in social anthropology at Clare College, Cambridge University, whose work focuses on citizenship through social movements and trade unions in Bolivia and Argentina. Lazar 2013 includes the article I cite, Lazar 2004.
4. She is referencing Polanyi (1957[1944]), who argued that society once embedded the economic, but with capitalism the economy was embedding society.
5. Guggenheim presented early results of the project at the Agrarian Studies seminar at Yale. He was then working for the World Bank, but is now senior social policy adviser for the AusAID–Indonesia Partnership Program and senior adviser to Afghan president Ashraf Ghani. Tania Li knows him, and so do I.
6. Tania Li is a professor of anthropology at the University of Toronto in Canada. She published an early version of the ideas considered here as Li 2006.

7. Anna Tsing is a professor of anthropology at the University of California at Santa Cruz. Tsing's piece was developed in a 2005 study group of Tania Li's. She says it grew out of *Friction* (Tsing 2009:151n3). The wild mushroom supply chain in *Mushroom at the End of the World*, in turn, was being researched while Tsing 2009 was being written.
8. Tsing (2009:155) uses the term "governmentality" for franchises, but she defines governmentality as rule by rule itself, citing Barry, Osborne, and Rose (1996). I would characterize franchises as government by discipline, and supply chains as governmentality.

References

Barry, Andrew, Thomas Osborne, and Nikolas Rose, eds. 1996 *Foucault and Political Reason: Liberalism, Neo-Liberalism and Rationalities of Government*. Chicago: University of Chicago Press.

Ferguson, James 2010 The Uses of Neoliberalism. *Antipode* 41(S1):166–84.

Foucault, Michel 2008 *The Birth of Biopolitics: Lectures at the College de France, 1978–1979*. Michel Senellart, ed. Graham Burchell, trans. New York: Picador.

Harvey, David 1990 From Fordism to Flexible Accumulation. Chapter 9 of *The Condition of Postmodernity*. Pp. 141–72. Cambridge, MA: Blackwell.

Lazar, Sian 2004. Education for Credit: Development as Citizenship Project in Bolivia. *Critique of Anthropology* 24(3) 301–19.

Lazar, Sian, ed. 2013 *The Anthropology of Citizenship: A Reader*. Malden, MA: Wiley-Blackwell.

Li, Tania 2006 Neo-Liberal Strategies of Government through Community: The Social Development Program of the World Bank in Indonesia. *International Law and Justice Working Papers* 2006/2. New York, NY: Institute for International Law and Justice, New York University School of Law.

Li, Tania 2007 *The Will to Improve: Governmentality, Development, and the Practice of Politics*. Durham, NC: Duke University Press.

Li, Tania 2014 Fixing Non-market Subjects: Governing Land and Population in the Global South. *Foucault Studies* 18:34–48.

Molyneux, Maxine 2002 Gender and the Silences of Social Capital: Lessons from Latin America. *Development and Change* 32(2):167–88.

Pettinger, Tejvan n.d. Structural Adjustment. *Economics Help*. www.economicshelp.org/blog/glossary/structural-adjustment/, accessed 10/20/2018.

Polanyi, Karl 1957[1944] *The Great Transformation: The Political and Economic Origins of Our Time*. Boston: Beacon Press.

Rose, Nikolas 1999 *Powers of Freedom: Reframing Political Thought*. Cambridge UK: Cambridge University Press.

Russo, Francesca, Jim Smyle, and Armando Guzman 2004 Community Driven Development Approaches in Natural Resources Management Projects: Lessons from Guatemala and Honduras. In *Thinking Out Loud V: Innovative Case Studies on Participatory Instruments*. Katherine Bain, Franka Braun, Indu John-Abraham and Monica Peñuela, eds. Report number: 30332. The World Bank.

Tsing, Anna L. 2005 *Friction: An Ethnography of Global Connection*. Princeton, NJ: Princeton University Press.

Tsing, Anna L. 2009 Supply Chains and the Human Condition. *Rethinking Marxism* 21(2): 148–76.

Tsing, Anna L. 2015 *Mushroom at the End of the World: On the Possibility of Life in Capitalist Ruins*. Princeton, NJ: Princeton University Press.

The World Bank Group 2004 Community Driven Development. www.worldbank.org/en/topic/communitydrivendevelopment, accessed 12/12/2019.

Wikipedia: The Free Encyclopedia n.d. Neoliberalism. https://en.wikipedia.org/wiki/Neoliberalism. Accessed 1/8/2020.

13

THE ECONOMY IN ETHNOGRAPHIES

Introduction

The economy that ethnographers observe is "on the ground" in households and communities. This is the place where what economists call "the market" or "the economy" meets livelihoods, and where neoliberal interventions are confronted by households and communities. In fact, it is a space targeted by neoliberal interventions, whether to create social or individual *homo economicus*, with the market or without it (see Chapter 12). These interventions—and increasingly conservation—are based on an economic perspective, which (1) does not question the existence of the market, and sees it as both inevitable and valuable; (2) assumes that local economies (domestic, self-provisioning, subsistence, especially in the developing world, and especially rural ones) are essentially separate from the market; and (3) sees these local economies as "poverty" and links them to environmental degradation. Development and conservation thus assume that stimulating economic growth—which typically means bringing those parts of the world they see as characterized by poverty and environmental degradation into the global economy—will solve poverty and environmental degradation. "Indigenous" communities will have their "free" choices channeled to things like REDD+ rather than farming or selling timber (see Chapter 12). This economic perspective needs to be questioned. Ethnographies do this.

This space where the market meets livelihoods is everywhere in the global South.[1] The ethnographies I have selected all involve agriculture, which is one of the main livelihoods in this space, though far from the only one. Conservation can be implicated in this meeting: in one of the ethnographies a park restricts the expansion of farming, cutting off options for farmers. Whenever conservation tries to use development to offset the livelihood losses its restrictions introduce to people, or simply to reward people for conserving, it is dabbling in this space. The environment is always affected as farmers change crops, add chemical inputs, or open new land. Understanding this space where the market and livelihoods meet is vital to effective conservation involving people. But it is hard to understand. The discipline of economics does not think well about this space.

I selected these ethnographies to stimulate us to think about "the economy" in new ways, in particular, in ways that are outside the discipline of economics. Stephen Gudeman and Alberto Rivera (1990) use ethnography to question the ability of economic theory to understand peasant economics, with a case study of farmers in Colombia that gives us a rare peasant household's perspective on the market, alongside peasant practices. The perspective opposes household and market; practices knit the two together. Tania Li (2014) attempts to explain the emergence of capitalist relations in the absence of external intervention, based on ethnographic research in Sulawesi, Indonesia, where swidden farmers themselves took the initiative to begin cultivating a cash crop, cacao, with the end result being that some accumulated land and some lost it. In this book Li uses Marxist theory, in contrast to her 2007 ethnography, *The Will to Improve*, which primarily used Foucault, so I will review it too. Timothy Mitchell (2002) is seeking to undermine the discursive power of "the market" on our thinking about economies, based on ethnographic fieldwork in Egypt, where a neoliberal market reform engineered by the IMF and the United States Agency for International Development (USAID) inspired peasants to return to self-provisioning. All of these books are based on ethnographic research, in Colombia, Egypt, and Sulawesi, Indonesia.

Thinking well about economies matters because economies *are* ecologies. In Aristotle's and Colombian peasants' terms, only the earth or the "land" is productive. Ecologies are our "base"; eating and selling both depend on ecologies. This is most apparent with agriculture, as in these ethnographies, but of course all production draws from and affects ecologies. Even with agriculture, we can't think narrowly: small-scale farming without chemical inputs draws on grasslands and forests as well as fields, as all agriculture using inputs draws on fossil fuels. Peasants know this.

A peasant critique of the market

Stephen Gudeman[2] has spent his career trying to understand peasant theories of the economy in order to put them forward as an alternative to the dominance of economics. His book, *Conversations in Colombia: The Domestic Economy in Life and Text* (1990), written with Alberto Rivera,[3] is structured as a "conversation" between Colombian peasants and (indented) theory, including Aristotle, dependency theory, and Marx. Gudeman and Rivera give us a Colombian peasant's "house" critique of "the market."

Gudeman and Rivera (1990) lay out the opposition between "house" and market from the peasant's perspective. Gudeman (1978) sees this relation as dialectical, meaning that it is an opposition that is also an interdependent relation. Gudeman

(2001) complicates this dialectic even further, seeing corporations as being like a peasant house in combining market and house practices, such that both corporations and peasant houses are hybrids of market and house thinking and practices.

The essence of the Colombian house critique of the market is the unfairness of market exchanges to the house. The market pushes houses, which are always on the margins, "beyond the frontier of profit" (Gudeman and Rivera 1990:155). The house is pushed to the margins because it "sells to buy" necessities it cannot store or does not produce (base → money → base′); the corporation "buys to sell," and the desired result is money itself, profit (money → base → money′) (Gudeman and Rivera 1990:143).

In this peasant critique, the house's "base" is threatened every time the household markets a cash crop: "One is losing the base little by little to the market" (Gudeman and Rivera 1990:143). The base is everything that allows a house to meet its needs and its ability to sustain itself in the future, including land, buildings, seed stock, and knowledges and practices. Houses freely lend seed, tools, food, labor, and even money to each other with no interest (Gudeman and Rivera 1990:144).[4] The house concept of a just price is defined by the base: a just price allows a house to meet its needs and its ability to sustain itself in the future (Gudeman and Rivera 1990:144).[5]

From the peasant perspective, the earth gives the base of the house:

> Over and over the people said, "the land gives" (*la tierra da*). After some time, we realized that the phrase was shorthand. It can be completed with the name of a crop, such as "the land gives maize." The land also gives "food," "a harvest," or "the crops," and only the earth "gives the base" of the house, meaning that all wealth comes from the land.
>
> *(Gudeman and Rivera 1990:25)*

Some land gives more "force" or produces certain crops better than others. Fertilizer can add force to the land, but chemical fertilizer, though it helps produce crops, "'burns the earth' and 'takes away' its force" (Gudeman and Rivera 1990:27). Everyone says that the earth gives less now (in the 1980s), that it is tired and needs to rest, that it needs more "help" (Gudeman and Rivera 1990:27). Because these peasants understand that only the earth gives, they also believe that their work "sustains the nation" (Gudeman and Rivera 1990:150).

In one of Gudeman and Rivera's examples, a farmer is doing well. Alfredo watches over the estate of an absentee landlord without salary, but in exchange for the use of the land to farm and pasture livestock. He practices slash-and-burn agriculture, raising potatoes up to three times before opening a new field. The old field is seeded with grass and used for pasture. The potatoes maintain his "house," and he has some surplus to sell. They are raised with few monetary costs. Guinea pigs, chickens, and piglets are primarily raised for house consumption. Alberto sells the larger timber that he cuts from the fields, cutting it into firewood and taking it on packhorse to the market town every Saturday. And he sells some cattle and sheep as they produce an "increase," but he only sells a few grown ones. He purchases foods at the market. There are other expenditures, for clothes, utensils, and tools, though the house tends to be thrifty with them. An unusual market opportunity explains why Alberto's house is doing well: a dairy was built outside this town, and Alfredo sends milk to it daily from his four milk cows. That milk covers all the monetary costs of the house, so the returns from surplus potatoes, small animals, firewood, and surplus cattle and sheep are used to increase the base. Without the dairy, Alfredo would be outside the margin of profit. He is now inside it, but he does not own the land he is using, and he is also converting it slowly from forest to pasture.

Another example is more typical. A woman says her house raises maize for eating and onions for sale. The maize, she said, had more house uses than the onions, for it maintains the people of the house, the hired workers, and animals. Onions had few house uses and more money costs. Money could be made on them only by feeding the workers maize and selling some maize to buy fertilizer for the onions, thus making savings and reducing the cost of the onions. So the low-expenditure maize supported the high-cost onions. "If you can't make savings using maize, you can't make money with onions. If one had to keep monetary accounts … or buy everything for the onions, there would be no gain" (Gudeman and Rivera 1990:158). In Gudeman and Rivera's analysis, her house is substituting "non-accountable" expenditures for "accountable" costs—and this is usual. Whenever they asked people about the domestic labor or foodstuffs that went into the cash crops, they would say: "We don't keep accounts of that" (Gudeman and Rivera 1990:159). They argue that these non-accountable expenditures diminish the base little by little over time—unless they get a "just price" for the onions that allows the house to replace its expenditures (Gudeman and Rivera 1990:159). In an important sense, then, the base could be said to be subsidizing the marketing of cash crops, and the base includes the land itself, the earth. A bit further down the road the maize itself will need chemical fertilizers.

Both of these examples provide a sense of how deftly peasants move back and forth between what we see as two distinct economies: production for profit and production for eating (subsistence, self-provisioning). Things outside their control, market opportunities and market prices, determine whether their houses are inside or outside the margin of profit. Gudeman and Rivera believe that most are outside, and their bases are diminishing over time. In both cases the land itself, the source of all force and wealth according to peasants, is diminished. They are struggling to survive. But things could be worse.

Cacao and capital formation

Tania Murray Li[6] has written about the onset of capitalism in two different field sites in the same province of Sulawesi, Indonesia, one with a history of interventions, about which she *wrote The Will to Improve* (2007), and one without such a history, about which she wrote *Land's End* (2014). *The Will to Improve* is centered on Lore Lindu National Park. Cacao[7] is the cash crop that brings capitalism in both areas. Capitalism, however, unfolds differently in the two sites. Li uses Foucault's governmentality and Marx to analyze the 2007 ethnography, governmentality being critical to understanding the history of interventions. She uses Marx through Robert Brenner and Ellen Wood to analyze the 2014 ethnography, which had no history of interventions.

Li (2007:20–1) writes, of the park site,[8] that government (using Foucault) and capitalism (using Marx) intersect: governmentality and capitalism together stimulate capital formation (or capital accumulation) and identity formation (or class formation of a sort). These processes are out of control of expert intervention, and not a natural outcome of market processes. Capital formation, often called capital accumulation, leads to agrarian differentiation through unequal control over land. This is the same in 2014 in the site with no history of interventions. The difference in Li's analysis of the park site is the addition of identity formation, which occurs because migrant Muslim and local Christian identities line up or overlap with land control and class, as migrants buy land from locals. This does not happen in the more isolated site, where differentiation occurs without class formation. In that site Li focuses on the development of capital relations.

The two sites had some similar factors affecting the process of capital formation or accumulation, and some critical differences. Cacao, whether adopted by migrants (2007) or locals (2014), commodified land that had not been previously commodified. Cacao is a tree, and planting trees solidifies claims to land, unlike other crops. Unequal access to capital to purchase land and cacao seedlings and unequal access to knowledge of its profitability led some people to sell off land in both sites. In the park site the capital and knowledge were in the hands of migrant Muslims (Li 2007:119); in the more isolated site they were in the hands of fellow Lauje. In both sites, forest exclusion in the 1970s for timber concessions and conservation meant that there was no longer any forest left in which the landless could open new swiddens. This set conditions for the emergence of a landless class in both places (Li 2007:106). Cacao has limited labor demands for the newly landless, which affected people in both sites. Conservation became the ultimate target of villagers' resistance in the park site (Li 2007:122). The Lauje in the more isolated site also considered that "land's end" (*Land's End* is title of Li's 2014 ethnography) was the critical factor, but no resistance emerged (Li 2014:17). Li (2007:97, 2014:7, 116) argues that the emergence of a landless class was not a "natural" outcome of market processes in either site.

The rest of my discussion focuses on Li's second and more isolated ethnographic site, because here the market meets livelihoods with no help from neoliberal "development." In this site, market relations are spreading all by themselves. Li (2014:6, 115) uses Marx through Robert Brenner and Ellen Wood, because they make the "shift from market-as-opportunity to market-as-compulsion the critical diagnostic of capitalist relations." It is compulsion that helps explain the erosion of previous relations that once guaranteed access to food, labor, land, and other sorts of help, and the emergence of capitalist relations (Li 2014:116). Capitalist relations are governed by competition and profit; they are an ensemble of relations characterized by private and unequal ownership of the means of production (land, capital) and non-owners compelled to sell their labor (Li 2014:8). Li (2014:116) writes that the emergence of capitalist relations needs to be explained: "The emergence of capitalist relations is not an inevitable progression or evolutionary unfolding." She describes the emergence as a conjuncture: "A conjuncture isn't a seamless whole. Its elements jangle and collide. They foreclose some pathways and open others" (Li 2014:149). The emergence of capitalist relations occurs "by stealth," "insidiously, piecemeal, and unannounced" (Li 2014:116, 148). It is the erosion of choice and the emergence of compulsion that distinguish capitalist relations (Li 2014:148).

The elements of the conjuncture that created the compulsion that defines capitalist relations include access to capital, individual ownership of land, the materiality of crops, market prices, the extraction of surplus value from wage-workers, and contingencies. Highlanders' autonomy was grounded in common land and an open land frontier; capitalist relations began to occur when the Lauje, in a context defined by timber concessions and parks, planted tree crops, which transformed commons into individual property, unequally (Li 2014:6–7):

> The process that dislodged them [indigenous highlanders] from their land wasn't initiated by land-grabbing corporations or state agencies.... It was the highlanders themselves who took the initiative to plant tree crops, which had the effect of individualizing their land rights and led to the formation of capitalist relations.
>
> *(Li 2014:3–4)*

Once land was limited, it wouldn't support food *and* cash crops, because it couldn't be fallowed (necessary to restore fertility to the thin, low fertility topsoil), so highlanders had reached "the point of compulsion: they no longer had a choice." They had to grow cacao (Li 2014:120). Competition meant that those growing cacao must seek profit to generate capital to continue to be owners, and their accumulation squeezed others out (Li 2014:8). The ability of some highlanders to prosper depended on the failure of their kin and neighbors (Li 2014:8).

Access to capital in this case study was usually wages from the biannual clove harvest on the coast or work as porters hauling timber, and thrift, including reducing consumption of coffee and cigarettes, reducing gambling, and reducing gifts to neighbors and kin (Li 2014:118). All examples of thrift are social practices typifying previous social relations, such that working hard for cash was insufficient; one had to begin cutting off social relations to continue to buy land and cacao.

Extraction of surplus value from wageworkers, who were neighbors and relatives, was also clearly a breakdown in social relations. Class relations emerged, cutting across kinship and neighborliness (Li 2014:140). Labor relations are particularly awkward.

Private ownership of land increased the effects of debt, because farmers could sell their land to pay it off. Similarly, the effects of contingencies like drought and illness were no longer random, because the richer could weather them but the poorer began a downward spiral, also often leading to land sales (Li 2014:135).

The ecology of the two cash crop options is also critical. Cacao doesn't yield for two to three years, and begins needing inputs (purchased fertilizer) between years 5 and 15 to continue to yield. If a farmer hasn't ended up with sufficient land to grow both food and cacao, in other words, he's going to have to purchase rice on the market while the cacao matures. Market prices, especially the ratio between the price of cacao and the price of rice, became critical. The alternative cash crop, clove, is also problematic: clove takes five or more years to yield, and provides a good harvest only in alternate years.

In sum, Li has given us ethnographies of the emergence of a market economy in two sites within the same province, both associated with the cash crop cacao. In both sites the first influx of cacao also privatized land for the first time. The Lore Lindu Park site had a long history of interventions, including conservation interventions. Cacao was first planted by coastal migrants, who were able to accumulate capital. As Islamic migrants became wealthier, a new sense of identity formed among the increasingly poor, Christian, locals. The more isolated Lauje site had no history of interventions, though timber concessions and a park had foreclosed options for expansion. The Lauje themselves started planting cacao, which led to the compulsion to keep buying more land and planting more cacao. Capitalist relations took hold, separating kin and neighbors. No identities formed.

Returning to self-provisioning

In southern Egypt in the 1990s, many rural households with small landholdings were growing increasing amounts of their own staple food, wheat, processing it at tiny village mills, and using it entirely for making bread for their families. They were also turning from tractors to wooden plows pulled by cows (Mitchell 2002:253–4). This is self-provisioning. What is surprising about this is its history: it is actually a return to self-provisioning after nearly 40 years of dependence on subsidized US wheat,[9] followed by IMF/USAID demands for neoliberal reform and US demands for market prices for its wheat in the 1980s (Mitchell 2002:248–50). In other words, Egyptian farmers returned to self-provisioning as a response to neoliberal, free market reforms. Their responses "moved away from or at cross-purposes with the logic of the market" (Mitchell 2002:254). In the 1990s more than two-thirds of the crop area in Egypt was planted in crops intended mostly for household use (Mitchell 2002:252).

As Timothy Mitchell[10] argues, an intensive system of food manufacture (i.e., self-provisioning—labor used to grow wheat, make bread, collect fuel and fodder, milk buffalo, raise chickens, etc.) is probably Egypt's largest industry. For the village and also probably for Egypt, the purpose of production for the market was to support the much larger system of self-provisioning (Mitchell 2002:255). The best example of production for the market is sugarcane. The Mahmud household had four acres, over two of which were used to grow wheat for household consumption. Another acre was used to grow sugarcane, their only cash crop. The rest of the land grew fodder and vegetables for the household. The reason for growing the cane was because they needed income to buy seeds, fuel for an irrigation pump, and fertilizer for household crops. It also provided credit, through a cash loan early in the year from the government sugar factory, without which they wouldn't survive the growing season. In other words, they needed income and the loan to buy inputs for household crops and food until the wheat ripened. This is the reverse of the pattern Gudeman and Rivera saw in Colombia, where the food crop was used to subsidize the cash crop.

This example also demonstrates that self-provisioning is not "pure." It does not create a separate, non-market sphere. Villagers without sufficient land to grow sugarcane worked without pay on large sugarcane farms in return for access to fodder, and sugarcane farmers gave loans for seeds and fertilizer to wheat farmers—because they depended on their free labor. This is certainly also true in Gudeman and Rivera's examples in Colombia. In Egypt in fact the market sphere and the self-provisioning sphere are mixed in most loaves of bread that village women bake, as they mix a little purchased white flour with their home-grown and village-milled wheat to make their own wheat last longer (Mitchell 2002:250).

Mitchell is able to document this response because he carried out ethnographic fieldwork. IMF and USAID saw national statistics showing that wheat production had increased as proof that their free market reforms had succeeded. They saw food manufacture as residual and insignificant according to the model of a market (Mitchell 2002:263). Mitchell suggests that "the market's place," not the system of local food production, is residual and insignificant.

Mitchell also critiques the idea of free markets, so close to the heart of neoliberalism. There is no free market in wheat, he argues: the marketing of wheat is controlled by five or six international grain-trading corporations, and both the US and the EU use systems of price supports, without which their wheat production would not be profitable (Mitchell 2002:257). In fact, he says, free market farming is impossible because it exposes farmers to price swings that they cannot survive (Mitchell 2002:263).

Finally, Mitchell (2002:248) argues that "the market" is not a singular, universal form. He sees the market, rather, as a discourse. He writes:

> We have to avoid the assumption that capitalism has an "is" and take more seriously the variations, disruptions, and dislocations that make

> each appearance of capitalism, despite the plans of the reformers, something different.
>
> *(Mitchell 2002:248)*

What is different here is "each appearance of capitalism," though he doesn't like the idea of multiple capitalisms (Mitchell 2002:248). It is "local complexities" and "existing networks and logics" that "resituate and transpose" the project of market reform (Mitchell 2002:261). The separate existence of either is only discursive, as is the opposition between them. In fact, Mitchell says that discursively this opposition "gives capitalism its identity" (Mitchell 2002:245). The opposition between "the economy" and "self-provisioning" is not real.

Conclusion

In Mitchell the market is only a discourse, or a project disrupted by local logics, and each local appearance of "capitalism" is different, due to those same logics. The market in Gudeman and Rivera is more singular, but its solidity stems from peasant discourse about it and what it is doing to them. Gudeman and Rivera's indented theory is also meant to show that our theory of the economy has a history that resembles peasant discourse. Both Mitchell and Gudeman have written many pieces meant to undermine the power of the discourse of economics. Though both Mitchell and Gudeman and Rivera base their pieces on ethnography, Li writes even closer to the ground, moving from theory to lists of factors or elements that all contribute to the emergence of capitalism and capitalist relations. Capitalist relations are not just a discourse to Li, but they must be explained wherever they emerge.

All these authors object to the opposition between "economy" or "market" or "capitalism" and self-provisioning, the "house," or previous relations. In Mitchell the opposition is only discursive, and "capitalism" relies on the opposition to establish its discursive universality. In reality, peasants in Egypt do not separate the two. In Gudeman and Rivera, peasants talk in terms of the opposition, but in reality they survive by mixing them. Something different happened in Li's more isolated research site: a point of compulsion was reached where those doing well with cacao had to keep buying land and planting cacao, and also a point of failure where those doing poorly were not left with enough land to grow food. Li argues that this point of compulsion appears slowly, as choices are cut off one by one, in a particular locality, by the disparate effects of things like conservation, price shifts, and the ecology of certain cash crops. It is not an "inevitable progression" or an "evolutionary unfolding." But the Lauje were not able to mix rice and cacao.

The readings in this chapter add to the complexity of the "economy" by embracing both eating and selling. Rice grown to eat is being displaced by cacao grown to sell in Sulawesi; Colombian farmers grow corn to eat with onions to sell, subsidizing the onions with the corn; in the rural areas of Egypt, wheat to eat is grown by some, and sugarcane to sell by others in the same villages. Sugarcane farmers depend on free labor from wheat farmers; wheat farmers subsidize their wheat with free fodder and loans from sugarcane farmers. And eating and selling don't stay apart: cacao farmers must now buy rice to eat as well as fertilizer, Colombian farmers sell corn to buy fertilizer for onions, and Egyptian wheat farmers labor for sugarcane farmers in exchange for fodder and loans to get them through to the next food harvest.

The movement of change, the "story" in these pieces, is quite different. In Mitchell small farmers respond to neoliberal reforms by returning to self-provisioning, a case that contradicts the usual free market wisdom (and escapes the notice of reformers). The system of food manufacture, in Egypt at that time, is more important than neoliberalism. In Gudeman and Rivera most peasants must farm both to eat and to sell (for necessities) even though they see the market gradually undermining the house's base. The market is eating up the houses on which it depends for subsidies. One farmer, however, thanks to a nearby dairy opening an opportunity, is doing well. In *Land's End* (Li 2014) capitalist relations are relentlessly, though complexly and gradually, displacing previous social relations, as opportunities become compulsions.

All of these things can and do happen, and what seem to be stories are only moments. We don't know what is happening in Egypt, Colombia, or Central Sulawesi now. What is clear is that farmers juggle eating and selling, that the market can open up opportunities for some, and that these opportunities can become compulsions that destroy some farming households. It is also clear that these understandings are only accessible ethnographically. Neoliberal economics does not see these things.

All these cases should temper our tendency to think that capitalism is singular, universal, inevitable, or good. In some places, at some times, a small "m" market or a particular market project, or a particular market crop, interacts with existing networks and logics or house economies or social relations in a way that encourages farmers to self-provision or makes it impossible for houses to profit, or creates landed and landless classes. All are possible, and there must be other possibilities as well.

All these cases should temper our enthusiasm for economic growth, and especially for interventions that draw people into "the market." In Egypt farmers protected what really mattered: food. In Colombia farmers chose to grow both food and cash crops, but weakened their households' economic bases in doing so. In Central Sulawesi some farmers chose to grow cash crops, but destroyed their neighbors and kin in doing so. The most secure people in these cases are the Egyptian farmers who were growing their own food. The most prosperous were those households in Sulawesi who decided early on to grow cacao (though the market for cacao is fairly volatile). The most desperate are the neighbors and kin of the most prosperous. Colombian farmers, in the middle, see themselves as being on a slow slide to poverty.

And what about the local ecologies that constitute the natural base of these economies? Mitchell doesn't tell us, but I like cows better than tractors, and local mills better than wheat imported from the US. Colombian farmers are well aware of the significance of soil, but have still adopted a cash crop that requires some fertilizer. In Sulawesi the shift is most massive, from swidden farming's fallowed forest to fertilized cacao plantations. If we scale out in Sulawesi, we can bring the cacao fields, the logged forest, and the protected forest in view. All together constitute Sulawesi's "appearance of capitalism."

I also think that the ecologies of particular crops matter a lot, along with their demands on labor and perhaps price stability for cash crops. This is clear with cacao, which just doesn't work alongside rice in thin tropical soils, has few labor demands, and had a very attractive price for a while. Wheat and sugarcane, in the same villages but not the same farms, seem to work well together. Corn and onions seem marginal. Farmers all over the world are trying to find combinations of crops that will allow them to eat and to sell. Perhaps we should be helping them.

Notes

1. Livelihoods actually confront the market everywhere, of course, not only in the global South. But interventions to bring people into the market tend to focus on the "developing" world (that term itself essentially means not yet in the modern market). This book's focus on interventions means that it is also focused on the global South.
2. Stephen Gudeman is probably the foremost economic anthropologist in the US now, and professor of anthropology at the University of Minnesota. He has a 1970 PhD in anthropology from the University of Cambridge and an MBA from Harvard. He began as a neoclassical economist, then moved to Marxism and dependency theory (Gudeman 1978), before turning to the close ethnography of folk models for economy.
3. Alberto Rivera, sometimes Rivera-Gutierrez, was Gudeman's doctoral student, and is now a practicing anthropologist in Guatemala.
4. Though Gudeman is thinking of individual households in this book, by 2001 he defines the base as inherently associated with community.
5. Compare this to the peasant view of extraction in Scott 1976.
6. Tania Li (2014:21) is an anthropology professor at the University of Toronto. She made nine separate visits to Sulawesi between 1990 and 2009, totaling about one year in duration, with none between 1998 and 2006. She never learned Lauje but used the same interpreter for each visit (27).
7. There was a cacao boom in Indonesia in the 1980s and 1990s. Indonesia's monetary crisis of 1997–8, especially the crash of the rupiah versus the dollar, greatly increased the price of cacao. By 2009 the market was poor, but it has since leveled off.
8. The park site is examined in more detail in Chapter 10.
9. US wheat was actually cut off during some periods when Egypt was not seen as an ally, returning when the two nations' interests once again converged. The nation of Egypt also did what it could to protect people against the global wheat market (Mitchell 2002:248–50).
10. Timothy Mitchell is a British political scientist and a professor of Middle Eastern studies at Columbia University. He is married to the anthropologist Lila Abu-Lughod. His most recent book is *Carbon Democracy*.

References

Gudeman, Stephen 1978 *The Demise of a Rural Economy.* London: Routledge and Kegan Paul.

Gudeman, Stephen 2001 *The Anthropology of Economy: Community, Market, and Culture.* Malden, MA: Blackwell Publishing.

Gudeman, Stephen and Alberto Rivera 1990 *Conversations in Colombia: The Domestic Economy in Life and Text.* Cambridge, UK: Cambridge University Press.

Li, Tania 2007 *The Will to Improve: Governmentality, Development, and the Practice of Politics.* Durham, NC: Duke University Press.

Li, Tania 2014 *Land's End: Capitalist Relations on an Indigenous Frontier.* Durham, NC: Duke University Press.

Mitchell, Timothy 2002 *Rule of Experts: Egypt, Techno-Politics, Modernity.* Berkeley, CA: University of California Press.

Mitchell, Timothy 2011 *Carbon Democracy: Political Power in the Age of Oil.* London: Verso.

Scott, James C. 1976 *The Moral Economy of the Peasant: Rebellion and Subsistence in Southeast Asia.* New Haven, CT: Yale University Press.

14

THE INVISIBILITY OF IMPLEMENTATION AND GOVERNMENTALITY

Introduction

Implementation, the actual implementation of a conservation project, can be seen as a singular and significant space for government and for policy.[1] It is the location where the discourses, economics, and politics of a project and its implementers meet local discourses and politics, and local economies and ecologies. Implementation is where conduct is supposed to be conducted (e.g., the space where user groups are formed, resources managed, the market entered), and conserving subjects are supposed to be cultivated. In other words, implementation is the site where governmental plans—whether by policy-making organizations or states—meet reality. It is the "witches brew" of practice that Foucault did not study. It is also the space where policy meets the outside world. Real conservation, the conservation that is written on real landscapes, happens here.

Implementation is also the space where metrology meets "the outside." Bruno Latour's (1987:251) metrology (see Chapter 7) is the enterprise to make of "the outside" a world inside which science and technology, facts and machines, and government and policies can survive. Latour's point is that these things can't survive unless they are able to spread "lab conditions" in front of them. This process could surely be studied for conservation, but my point here is that "the outside" is the space of implementation.[2] As Latour (1997:250) says, "As soon as the outside is really encountered, complete chaos ensues." Using the example of invading Russian forces trying to follow maps of Czechoslovakia when all the street signs had been torn down, he says: "When the out-thereness is really encountered, when things out there are seen for the *first* time, this is the end of science, since the essential cause of scientific superiority has vanished" (Latour 1977:254). The street signs had extended lab conditions, making the maps work; without them all the science and knowledge of mapping didn't work.

Tania Li (2007:27–8) uses comparable language to explain her research strategy, which brings together an analysis of governmental development interventions with an analysis of "what happens when those interventions become entangled with the processes they would regulate and improve":

> This strategy takes me beyond the plan, the map, and the administrative apparatus, into conjunctures where attempts to achieve the "right disposition of things" encounter—and produce—a witches' brew of processes, practices, and struggles that exceed their scope.

Li (2007:28) argues that both the "program" and the "'beyond' of programming," "the messy consequences" of programs, are real and merit attention from ethnographers.

How does policy succeed in "the outside" or the "beyond of programming" that we call implementation? The space of implementation is also one of the spaces of ethnography,[3] and ethnographies of projects are the best sources of understanding about it. As David Mosse (2004:641) says, "My concern is not whether, but *how* development works. The approach is *ethnographic*." I will consider three answers to this question below. All three are grounded in ethnography, though in Scott's case he is analyzing ethnographies of others. The first is representations of implementation success in terms of the policy, which can be made and supported by both implementers and local people. This answer comes from David Mosse, whose 2005 book, *Cultivating Development*, analyzes over 12 years of project implementation. The second answer is informal knowledge, of implementers and local people, as argued by James C. Scott. The third answer is compromises between implementers and local people, as analyzed by Tania Li, and also based on ethnographic research of a project. Mosse and Li both emphasize "how" rather than "why" questions: How is policy success produced? How is government accomplished?[4]

Considering its significance, the singular and significant space of implementation is remarkably invisible to the policy-makers and the states or organizations behind them who have promoted the governing discourse (e.g., conservation) and planned the desired conducts and subject formation. David Mosse argues that the practices of implementation are in a black box for policy-makers. James Scott argues that the informal knowledge necessary to implement state plans is devalued; states dream of excluding it. Tania Li argues that the accomplishment of rule depends on compromises in implementation; as we will see, invisibility is implied in the word "compromise." Implementation does not change the discourses that animate policy, or the models (like community-based natural resource management) or practices (like user group formation or trainings) that policy-makers return to again and again in project plans. We might say that there is an anti-implementation machine parallel to Ferguson's anti-politics machine (see Chapter 3). This invisibility is not ignorance; it is the discursive power to erase at work. Project planning processes exclude the realities of project implementation.

This means that the space where government meets reality is relatively cut off from government. Implementers (whether international consultants or local-level officials) are relatively free to act outside the box of project plans (though thinking outside the discourse will continue to be challenging). This is reason for optimism: the actual effects of conservation occur in this invisible space, full of politics and power, but also full of freedom.

Local peoples are also usually free. First, they are free because we are talking about liberal forms of government not based on coercion (with the exception of coercive forms of conservation in parks). Second, local people have their own conducts, personhoods, politics, economics and economies, knowledges, and discourses; in short, their own cultures and societies.[5]

The invisibility and freedom that characterize the space of implementation may typify the type of rule Foucault called neoliberal governmentality—the sort of rule that does not want to rule too much. The singular and significant but invisible space of implementation is built into the structure of development and conservation interventions as a space that doesn't have to be directly ruled. Organizations like the World Bank and USAID, for example, don't "do" implementation: they contract it out, managing the contractors. Yet Tania Li argues, and I agree, that this is how rule is accomplished, and how policy "succeeds."

The invisibility of implementation

David Mosse[6] (2004:641) argues that implementation is in a "black box of unknowing," a space that is invisible, concealed, and/or devalued by development and conservation players. There is a gap "between the '"monotheistic privilege" of dominant policy models' and the '"polytheism" of scattered practices' surviving below" (Mosse 2004:645 quoting de Certeau 1984:48).

This fact leads him to five propositions, illustrated by his own experience of over 12 years with the Indo-British Rainfed Farming Project, which has participation as its primary policy. First, policy[7] functions to mobilize and maintain political support rather than to direct implementation (Mosse 2004:648). In his project, "participation," "a single, technical-rational, politically acceptable, ambitious and ambiguous project model" translated "a multitude of contradictory interests and cross purposes" (Mosse 2004:651). Participation worked well "to legitimize, mobilize support, and bring people and agencies together across organizations, nations, and cultures (by taking on more and more agendas, increasing complexity, burying differences)," but did not work to provide a good guide to action, practice, or implementation (Mosse 2004:651).

Second, the actual practice of implementation is thus driven not by policy but "*by the exigencies of organizations and the need to maintain relationships*" (Mosse

2004:651, author's italics). In this project the cooperating Indian organization was a manufacturing and marketing giant, whose managerial regime and organizational systems and procedures "prioritized familiar conventional programmes," resulting in Participatory Rural Appraisals, community problem analyses, and village workplans that produced a fixed set of interventions (e.g., crop trials, tree nurseries) unresponsive to local needs (Mosse 2004:652). Women and really poor people were also excluded (Mosse 2004:652). And local people were "unruly objects of development" who:

> Strive to be modern when we want them to be indigenous, chaotic when we demand order; they present themselves as our clients and employees when we call them partners; dependent when we insist on their autonomy. They make a mockery of our models and our explanations.
>
> *(Mosse 2004:654)*

Mosse's third proposition is that project success is produced in representations that go upwards and outwards to secure reputations and funding, and that must try to support policy coherence, which is fragile (Mosse 2004:654–5).[8] Mosse is looking at the work that people in development and conservation must do to create persuasive, coherent narratives—narratives that claim success in terms of the policy being implemented—out of the complex realities and multiple agendas of project implementation. Implementers work to support policy coherence,[9] to maintain the discursive power, in this case, of the policy of participation. As Mosse explains:

> While the coherence of design unravels in the practical unfolding of a project, everybody is particularly concerned with making, protecting, elaborating and promoting models with the power to organize authoritative interpretations, concealing operational realities, re-enforcing given models and limiting institutional learning.
>
> *(Mosse 2004:664)*

This is the work of "translation," "creating order and unity through political acts of *composition*" (Mosse 2004:647). The ethnographic task:

> Involves examining the way in which heterogeneous entities—people, ideas, interests, events and objects (seeds, engineered structures, pumps, vehicles, computers, fax machines, or data bases)—are tied together by translation of one kind or another into the material and conceptual order of a successful project.
>
> *(Mosse 2004:647)*

Note that the objects—the seeds, engineered structures, etc.—colonize "the outside" with lab conditions, as do the standardized programs of participation and the fixed set of interventions mentioned above. Mosse (2004:657) is not accusing anyone of duplicity: "Interpreting and presenting events through the official model was a habit (of mine too)." And it's not just implementers who do this work; it is in the interests of all subordinate actors in development to do this work: "tribal villagers, fieldworkers, office staff, even project managers and their bosses in relation to donors" (Mosse 2004:665).

Policy-makers do this work of representation too. As Mosse says:

> The operational control which bureaucracies or NGOs have over events and practices in development is always constrained and often quite limited. What is usually more urgent and more practical is control over the *interpretation* of events.
>
> *(Mosse 2004:646, author's italics)*

Even that control is fragile, because it depends on being sustained by an interpretive community (Mosse 2004:646). Drawing on Bruno Latour, Mosse, says that:

> the success of policy ideas or project designs is not inherent (not given at the outset) but arises from their "ability to continue recruiting support and so impose ... [their] growing coherence on those who argue about them or oppose them."
>
> *(Mosse 2004:646, quoting Latour 1996:78)*

Policy coherence is thus created and sustained socially, and implementers play an important part in this. As Mosse concludes, "The coherence and order of a successful project is always vulnerable; interpretations can fail" (Mosse 2004:665).

Mosse's (2004:658, author's italics) fourth proposition is that "*projects do not fail; they are failed by wider networks of support and validation.*" His own project failed due to a "*crisis of representation* in which external policy shifts had made 'participation' a less than adequate metaphor for the development process" (Mosse 2004:658, author's italics). As he explains: "Here was a project that had lost its interpretive community and its context, whose networks were in tatters; a project whose old supporters, advocates and project-policy brokers had moved on" (Mosse 2004:660).

His fifth and final proposition is that success and failure are thus policy-oriented judgments that obscure on-the-ground project effects, including positive ones (Mosse 2004:662). His project created "new avenues of non-state patronage which were advantageous in a remote tribal area, providing new input lines for improved technology or marketing possibilities" (Mosse 2004:662).

Cultivating Development is probably the single most important piece David Mosse has written, and was very controversial. In the preface, he writes that:

> A number of key actors (those in managerial positions) took strong exception to my account.... They disagreed fundamentally with my conclusions ... I held to the truthfulness of my analysis, but offered to record alternative points of view in a postscript. My refusal to suspend publication ... provoked written complaints to my academic managers, my university's research ethics committee, the Chair of my anthropological association ... and my publisher.
>
> *(Mosse 2005:ix)*

This controversy suggests that policy-makers ("those in managerial positions") believe in the implementability of policy, in short that there is a discourse of policy as well as development and participation, and that that discourse cannot tolerate the idea that implementation might not really reflect policy at all.

Mosse takes responsibility "for the naivety, over-ambition, ignorance and wrong-headedness" of his own contributions to the participatory project. He also argues that "ethnographic writing opens a rift between different epistemologies," "to reflect and to write means striving to break free from ... the discursive hold of even one's own cherished policy discourse" (Mosse 2005:x–xi). Mosse spent 12 years writing reports that helped strengthen a policy discourse, but was eventually able to step outside of this and think and write about the work that implementers and locals, including Mosse himself, did to maintain the power of the policy discourse of participation. This is why Mosse champions the ethnographic approach.[10]

The invisibility of mētis and state power

"Mētis" is James C. Scott's[11] (1998) term for the rich, practical knowledge required for any complex activity: "know-how ... common sense, experience, a knack," "cunning intelligence," a "touch" (Scott 1998:311, 313, 320). Mētis is vernacular, local, and situational, and may be "keyed to common features of the local ecosystem" (Scott 1998:312, 316). It is immersed in practice and experience; it is what you can only learn by imitating and practicing (Scott 1998:314–15). It is improvisational and adaptable, and most necessary in settings that are complex and mutable (Scott 1998:309, 316, 320). It depends on "exceptionally close and astute observation of the environment" (Scott 1998:324). Mētis is not "traditional knowledge" (Scott 1998:331). Vis-à-vis state plans it may be illegal; vis-à-vis the formal economy it is often informal (Scott 1998:309). "Broadly understood, mētis represents a wide array of practical skills and acquired intelligence in responding to a constantly changing natural and human environment" (Scott 1998:313). It is the opposite of "thin" state simplifications. Mosse (2004:653) argues that it is essential for development project implementation.

One of the nicest examples of mētis that Scott (1998:6) gives is the work-to-rule strike, in which workers virtually halt production simply by following the rules meticulously—that is, by withdrawing their informal know-how. Another example is everything that was excluded from the *normalbaüme* (see Chapter 7), which included the social uses of the forest and the forest as ecology, without which the forest died. Social uses and ecology, like mētis, are informal and devalued, and it is mētis, not forest science, that knew them. Another example is the dependence of the modern science of developing pest-resistant hybrid seeds on germ plasm from rare developing-world cultivars (Scott 1998:270).

Modern science and technology are valued: "The spheres of human endeavor that are freest of contingency, guesswork, context, desire, and personal experience—and thus free of mētis … came to be perceived as man's highest pursuits" (Scott 1998:321). Mētis is devalued: "A certain understanding of science, modernity, and development has so successfully structured the dominant discourse that all other kinds of knowledge are regarded as backward, static traditions" (Scott 1998:331). He mentions "Third World developers, and World Bank officials" in particular. The practice of policy-making may be more like mētis than it knows. Scott cites Lindblom's depiction of actual policy practice as "the science of muddling through," a description that Mosse would second (Scott 1998: 327–8).

Modern states dream of excluding mētis: "The destruction of mētis and its replacement by standardized formulas legible only from the center is virtually inscribed in the activities of both the state and large-scale bureaucratic capitalism" (Scott 1998:335). However, as Scott argues, modern states and mētis are locked into a relationship with each other; however devalued, mētis is essential to the modern world. "Formal order," he writes, "is always and to some considerable degree parasitic on informal processes, which the formal scheme does not recognize, [but] without which it could not exist" (Scott 1998:310). Mosse (2004:653) cites this same passage to suggest that his development project was not manageable without these same informal processes. Scott (1998:425–6n20) cites Latour as an introduction to the literature studying actual scientific practice versus its codified form in things like publications. He says that "any formula that excludes or suppresses the experience, knowledge, and adaptability of mētis risks incoherence and failure" (Scott 1998:319).

Mētis is threatened; the combination of the "universalist pretentions" of science and technology and authoritarian state power are "truly dangerous to us and to our environment" (Scott 1998:340–1). But it cannot disappear altogether, for when modernist schemes "come close to achieving their impossible dreams of ignoring or suppressing mētis … they all but guarantee their own practical failure" (Scott 1998:340). Thus, state officials charged with managing some of the most extreme high-modernist plans, as Scott says, "as often as not … recoiled before the abyss and came to tolerate, if not condone, a host of informal practices that in fact underwrote the survival of the official scheme" (Scott 1998:310). On the whole, Scott (1998:7) sees social and natural diversity as resilient; state schemes don't get to destroy them in the end. This is where mētis comes into his ideas. And this anticipates Tania Li's compromises.

Compromises and state power

The term "compromise," as Tania Li[12] explains, is "an agreement between two parties," in which "agency is distributed, if unevenly: both sides have a 'power to'." It also implies "a level of conscious knowledge and understanding of what is being gained and given up. It thus carries with it the sense of betraying or compromising oneself" (Li 1999:298).[13] This, Li says, is the normal position of subjects complying with rule. She extends this position to include state functionaries, who are "similarly obliged to make compromises and live lies." Then she extends it further to "relations of rule," especially development, which "can also be compromised, or put at risk" in encounters with the ruled (Li 1999:299). Everyday compromises are Li's (1999:296) answer to the question (a "how" question): "How is rule accomplished?"

Compromises are cultural relations: they depend on shared cultural knowledge. They are "culturally intimate—but often uncomfortable—forms of engagement" (Li 1999:315). Thus "relations of rule are cultural relations" too. Everyday compromises, and thus rule, are worked out in the "contingent and compromised space of cultural intimacy" (Li 1999:295). In development, this compromised space lies at the "interface between development projects and those they target" (Li 1999:298). Compromises are cultural, not "engineered by an omniscient and very subtle state for the purpose of rule," nor "the ad hoc invention of strategic actors" (Li 1999:315).

Li is writing about a 50-year-old program for the resettlement of "primitive" people ("*masyarakat terasing*"), begun by the Dutch colonial office and continuing in New Order Indonesia. The program is prone to failure (typically the abandonment of a resettlement site), and has been criticized by journalists and NGOs, but it persists. It is thus a good example of how difficult it can be to govern, and how essential everyday compromises are to state power. What opens a space for the compromises to occur is the generic vagueness of the target population, "*masyarakat terasing*." This vagueness gives "room for maneuver" on both sides. Li (1999:304) argues that this maneuvering space "does not indicate the absence or weakness of rule but signals, rather, one of the ways in which it is accomplished." In Mosse's terms, it means that interpretation becomes more important than reality.

Li visited project sites in Sulawesi and Kalimantan. In Sulawesi, half of the people to move into the resettlement village were actually mountain people; the other half were Muslim coastal villagers, who no one saw as primitive. Within a couple of years most of the mountain people had moved back home. Officials agreed to let more coastal villagers take their place. In the Kalimantan site, the Punan were the people most often seen as primitive, but most were excluded from the project as "too primitive." Four Punan families (out of 75) moved into the site, but soon left. The site ended up being filled largely from a nearby village significant enough to already have a primary and secondary school. But the project was a success:

> The houses had been built ... the recipients had recognized their subject position, confirmed their exotic origins, and shown evidence of a (presumed to be new) capacity to work on voluntary collective projects ... they were in place (had not run away); and when needed to greet official visitors and show evidence of deference and gratitude, they were there on command.
>
> *(Li 1999:307–8)*

Li's compromises are similar to Scott's informal practices, mētis,[14] though he is talking about devalued knowledge necessary for state projects, while she is talking about cultural relations between implementing officials and target peoples necessary to the accomplishment of rule. Everyday compromises between state functionaries and citizens are as essential to the accomplishment of state rule as mētis is, because the state would be even more vulnerable to exposure without everyday compromises between its functionaries and citizens (Li 1999:316). Compromises actually require mētis; they depend on people's "intimate knowledge of their own state system, which includes the knowledge of 'how to go on' in a variety of contexts, including when up against a problematic plan or rule" (Li 1999:315).

From the perspective of citizens, Li sees rule as an accomplishment whose implementation involves compromises that give them room to maneuver, and, thus, within rule, some agency (Li 1999:316). This resembles Scott's resilient social diversity, but I believe Li sees states as more fragile than Scott does. "Categories that manifestly do not fit, plans that fail, and compliance withheld or withdrawn," she writes, "expose the fragile nature ... of the very idea of 'the state' as knower, arbiter, and provider for 'the people'" (Li 1999:297). Compliance, in particular, is a necessary and difficult accomplishment, signaled by visiting officials' first question: "Are the people here?" (Li 1999:298, 307). Development in Indonesia is also peculiarly fragile, in part due to "the anger of those whose land, forest, and other means of livelihood are appropriated for state or private schemes" (Li 1999:300).

Though Li does not characterize compromises as invisible, it is clear that they would not support state rule if they were visible. In fact, the invisibility is implied in the word "compromise": that sense of betraying oneself. Part of the compromise is that the non-primitive targets are not going to signal their inappropriate position, and the visiting official is not going to probe into it either.

Conclusion

In this chapter I have put two pieces talking about state power side by side with a piece on policy implementation in development. From my perspective, both states and development organizations—and conservation organizations—are governing. What I hope I have suggested is that the odd, invisible space of implementation—characterized by mētis and compromises—is essential to governing. Most of the realities of implementation do not reach the bureaucracies of policy and state, whose politics depend on interpretations. State rule is accomplished, and policy interventions "succeed," in interpretation. Whether we characterize this as governmentality or not, it seems to me to describe much of conservation as well as development.

There are good and bad implications to this. As I have said, the singular and significant space of implementation gives reason for optimism to local peoples, and also to some implementers. In Li's terms, it gives them some room to maneuver. Governmentality, after all, is based on the freedom of those governed—and this is one of the ways in which they (and we) are free. For conservationists, however, the invisibility of implementation seems troubling. For the environment is invisible too. People abandoned resettlement sites in the project described by Li because the soil was not suitable for the sort of agriculture imagined by the project, a fact interpreted away by compromises for a time. But ecology ultimately dictated that farming had to occur elsewhere, or differently, if the people were to survive. Those compromises had effects on the land, too, likely degrading poor rainforest soil and adding some chemical fertilizers and insecticides—and those effects are also invisible.

The effects of development on the environment are certainly invisible, but the effects of conservation itself are for the most part also invisible. In Chapter 1, I noted that it was impossible to figure out if conservation was succeeding; all one finds is a literature about how little data evaluating conservation interventions is being collected. This fact means that the implementation of conservation is invisible.

Both Mosse and Li based their ethnographic research on "how" questions, and Mosse championed the ethnographic approach. Their questions—How is policy success produced? How is government accomplished?—can only be answered through ethnographies. As David Mosse (2004:641) says, "My concern is not whether, but *how* development works. The approach is *ethnographic*."

Notes

1. The term "implementation" is nowadays used for processes that are not "on the ground" in the ethnographic sense, but occur at national, regional, and global levels, e.g., the implementation of the sustainable development goals. Mosse (2004:641) argues that this only increases the "black box of unknowing." In Chapter 16 we will see that it is possible to do ethnographic research on these processes too.
2. Conservation implementation probably rarely occurs nowadays in a space without a history of conservation implementations, which means that some "lab conditions" will already exist in the space.
3. As is policy programming itself.
4. Foucault also asks "how," not "why," questions (see Chapter 1).
5. Implementers will see local peoples through a haze of false discourses about things like "community," "subsistence," "poverty," and nature-degrading, to keep the list short.
6. David Mosse is a British anthropologist, and professor of social anthropology at the School of Oriental and African Studies, University of London. Along with *Cultivating*

Development, he has published several other edited books on development, two of them with David Lewis.

7. By policy, Mosse (2004:648) means models, strategies, and project designs.
8. A similar ethnography of a policy model influenced by Mosse is Rap (2006), who argues that the success of a policy model is only a success within a policy network, and that it is part of a cultural performance rather than based on evidence. Thanks to Zander Pellegrino for this suggestion.
9. By "policy coherence," Mosse (2004:647n13) means "the articulation of different, distinct elements which can be rearticulated in different ways because they have no necessary 'belongingness'" (citing Li 2000:152 citing Hall 1996:141–2)
10. Mosse reviews the state of theory about development in the 1990s, promoting an ethnographic approach over instrumental or critical ones. He links Foucault to critical approaches, e.g., the "relentless Foucauldian micro-physics of power" in Ferguson. I think he is mistaken about both Foucault and Ferguson. He also links Foucault to ethnographic approaches, saying that Tania Li has used Foucault's notion of governmentality "to show how policy regulates social life and makes subjects … not by repression … but through a productive power which engenders subjectivities and aspirations" (Mosse 2004:644).
11. James Scott is an American political scientist and anthropologist by adoption, probably because he has championed qualitative research over quantitative. He has been teaching at Yale since 1976 as the Sterling Professor at Yale and director of the Agrarian Studies program; before that he taught at the University of Wisconsin–Madison.
12. Tania Li is a professor of anthropology at the University of Toronto in Canada.
13. Mosse (2004:645) cites Li 1999 as demonstrating that development schemes require compromise.
14. Li 1999 does not cite Scott 1998.

References

de Certeau, Michel 1984 *The Practice of Everyday Life*. Steven Rendall, trans. Berkeley: University of California Press.

Hall, Stuart 1996 On Postmodernism and Articulation: An Interview with Stuart Hall. In: *Stuart Hall: Critical Dialogues in Cultural Studies.* D. Morley and K.H. Chen, eds. Pp. 131–50. London: Routledge.

Latour, Bruno 1987 *Science in Action: How to Follow Scientists and Engineers through Society.* Cambridge, MS: Harvard University Press.

Latour, Bruno 1996 *Aramis, or the Love of Technology.* Catherine Porter, trans. Cambridge, MA: Harvard University Press.

Li, Tania M. 1999 Compromising Power: Development, Culture, and Rule in Indonesia. *Cultural Anthropology* 14(3): 295–322.

Li, Tania M. 2000 Articulating Indigenous Identity in Indonesia: Resource Politics and the Tribal Slot. *Comparative Studies in Society and History* 42(1):149–79.

Li, Tania M. 2007 *The Will to Improve: Governmentality, Development, and the Practice of Politics.* Durham, NC: Duke University Press.

Mosse, David 2004 Is Good Policy Unimplementable? Reflections on the Ethnography of Aid Policy and Practice. *Development and Change* 35(4): 639–71.

Mosse, David 2005 *Cultivating Development: An Ethnography of Aid Policy and Practice.* London: Pluto Press.

Rap, Edwin 2006 The Success of a Policy Model: Irrigation Management Transfer in Mexico. *Journal of Development Studies* 42(8):1301–24.

Scott, James C. 1998 *Seeing Like a State.* New Haven, CT: Yale University Press.

15

PRACTICES OF ASSEMBLAGE AND ASSEMBLAGES OF EFFECTS

Introduction

The term "assemblage" is generally used in anthropology to mean assemblages deliberately constructed by people or groups of people, sometimes in competition with others, for a specific purpose. The assemblage always includes people, and often also includes discourses. It may also include "things": specific technologies or their components. I will use Tania Li's (2007) emphasis on "practices of assemblage" to analyze this sort of assemblage. She draws on the ethnographic example from joint forest management; its elements include things like trees, socially situated subjects like villagers and scientists, objectives like profit and control, and knowledges, discourses, institutions, laws, and regulatory regimes (Li 2007:266). I will also use examples from Timothy Mitchell (2005, 2008) to illustrate these practices of assemblage. In Li (2007:263) assemblage elements include "discourses, institutions, forms of expertise and social groups whose deficiencies need to be corrected, among others." In Mitchell (2008:1117), the elements in his example of Edison's electrical project include:

> not only ties among investors, politicians, and technicians, but also circuits for the transmission of capital … generating stations to transform coal into electric power, carbon filaments whose resistance was calibrated to the current-carrying capacity of copper cables and to the cost of the copper, a system of patents … and cable networks.

I argue that when these assemblages are successful and an intervention occurs, the *effects* of the intervention, typically unplanned and unanticipated, also constitute an assemblage: effects are interrelated, though the connections between them may defy explanation. I will use Mitchell (2002) to illustrate assemblages of effects. In this historical ethnography, the building of a dam in Egypt generates enormous chains of effects, which Mitchell describes as an "amalgam of human and non-human, things and ideas," and which I see as an assemblage of effects. Acknowledging that heterogeneous things shape one another is important to conservation, in which the interactions of things like existing ecologies and economies with expert plans is so often a surprise.

Ecology plays a very important role in assemblages of effects: in Egypt, for example, increasing the height of the Aswan dam in 1933 allowed year-round irrigation in most of the country and allowed islands of curly pond weed to grow in the Nile, where the hydraulic power of the water had been slowed. The chain effects were to open up new land for mosquitos and give them a new way to get to Egypt. By 1942 a malaria epidemic (to which the ecology of the *Anopheles* mosquito and *Plasmodium* parasite contributed) killed one to two hundred thousand people. Another effect of irrigation was to increase the need for fertilizers, because land was no longer getting nutrients from the annual flooding of the Nile. Nitrate fertilizer was supplied by German factories, and was suddenly cut off after 1939—because nitrates make both fertilizers and explosives. This led to a shortage of food that also took many lives, and left people too weak to survive malaria. Quinine was not available due to the invasion of Java, the source of quinine, by Japan. A war against mosquitos ensued led by a Rockefeller Foundation expert, who used DDT in 1943; we now know that DDT breaks down very slowly, poisons insects and fish, and harms plants and animals. This is an assemblage of interconnected elements that all came together in Egypt's 1942–4 disaster (and I left out the war, which killed fifty to seventy thousand soldiers in one 1942 battle, and left the countryside littered with land mines) (Mitchell 2002). Assemblages include people and ecologies; in fact, they may be more useful to think about environment–human relations than ecosystems with human beings in them. Conservation cannot be practiced without ecological and social assemblages and assemblages of effects. Assemblages may be a useful way to think about what we are really doing.

The idea of an "assemblage" can be traced to Foucault, Gilles Deleuze and Félix Guattari, and Bruno Latour, with Tania Li using Foucault and Deleuze and Guattari, and Timothy Mitchell using Latour. In Foucault, "assemblage" is sometimes a synonym for "apparatus" (*dispositif*), from which Ferguson's "machine" is derived. Here is a definition of apparatus from Foucault (1980:194):

> A thoroughly heterogeneous ensemble consisting of discourses, institutions, architectural forms, regulatory decisions, laws, administrative measures, scientific statements, philosophical, moral and philanthropic propositions—in short, the said as much as the unsaid.... The apparatus itself is the system of relations that can be established between these elements.

Foucault (1980:195) continues his definition to say that "there is a sort of interplay of shifts of position and modifications of function" between the elements of an apparatus. I will return in the conclusion to this chapter to Foucault's concept. Tania Li (2007:264) begins her discussion with Foucault, citing this same passage.

Li introduces Deleuze and Guattari to help her talk about the potential of assemblage "to finesse questions of agency by recognizing the situated subjects who do the work of pulling together disparate elements without attributing to them a master-mind or a totalizing plan," saying that the "diffusion of agency" is usefully elaborated by them (Li 2007:265). Deleuze and Guattari (1987:88, authors' italics) describe the tetravalence of assemblages, which includes a horizontal axis with two segments, one of content ("a *machinic assemblage* of bodies, of actions and passions, an intermingling of bodies reacting to one another") and the other of expression ("a *collective assemblage of enunciation*, of acts and statements"), and a vertical axis with "*territorial sides*" (which stabilize the assemblage), and "*cutting edges of deterritorialization*" (which destroy it). They give the feudal assemblage as an example. The machinic assemblage includes the body of the earth and the social body; the bodies of the overlord, vassal, and serf; knight, horse, and "their new relation to the stirrup"; and weapons and tools (Deleuze and Guattari 1987:89). The collective assemblage of enunciation includes the "juridical regime of heraldry" and "oaths and their variables" (Deleuze and Guattari 1987:89). On the other axis are simply feudal territorialities and reterritorializations, and the deterritorialization that carries it all away. Then, they say, "We would have to consider how this combines in the Crusades" (Deleuze and Guattari 1987:89).[1]

Latour's actor-networks are also assemblages (see Chapter 7). Actor-networks are constructed, and include people and non-people in comparable positions. Metrology is the "progressive extension of a network" "*outside* of the networks that gave birth to them" (Latour 1987:248–9). Mitchell uses the metrological actor-networks. Thus, Edison is from the beginning invested in an electrical system that will use light bulbs. Without that, the bulb has no place outside the lab that developed it. In Mitchell, assemblages are metrologies.

Systems have recently been defined as "groups or combinations of interrelated, interdependent, or interacting elements forming collective entities" (Arnold and Wade 2015:675). None of these assemblages are systems in the sense of forming a "collective entity" or whole that is logical, consistent, stable, unified, or integrated. Even deliberately constructed assemblages are not controlled by anyone. The main thing assemblages and systems have in common is an emphasis on relations between elements and some openness in what constitutes an element.

Both authors ask "how" questions. Li asks how an assemblage is assembled, and Mitchell how the modern economy is produced. Li (2007:264) says that "the answers to 'how' questions … deserve a more central place in studies of government than is currently accorded." These questions concern the *practice* of assembling, and can only be answered ethnographically. They include: how is contestation manifested? And given their fragility, how are assemblages secured, i.e., what practices secure them?

There is nothing trivial about assemblages. In Li practicing assemblages is also practicing government in the sense of governmentality. In Mitchell assemblages "bring the economy into being." Assemblage thinking is how we govern the world, and it could help us to conceive of the effects of our governing before they destroy it.

Practices of assemblage

Tania Murray Li[2] (2007:263) argues that although interventions are shaped by governmentality, they should be analyzed as practices of assemblage, "the on-going labor of bringing disparate elements together and forging connections between them." Practices of assemblage, she says, fill the gap between governmentality's will to improve and the things that make government so difficult to accomplish (Li 2007:263). This is another analysis of a "how" question: how is government accomplished (in Chapter 14), how is an assemblage assembled?

Assembling is hard work, as demonstrated by Li's list of six practices: (1) forging alignments; (2) rendering technical; (3) authorizing knowledge: assumptions must be confirmed and critiques contained; (4) managing failures and contradictions, often by devising compromises (see Chapter 13); (5) containing politics without infringing on participants' freedom; and (6) reassembling when the ground shifts, as it inevitably does. All six practices necessary to assembling are challenging.

Li illustrates the practices of assemblage through an assemblage that is fraught with tensions, but has nevertheless persisted for nearly 40 years: community forest management (CFM). Practices include, first, forging alignments, especially with villagers, since this intervention carries the word "community." This is particularly difficult because of a "line of fracture" between forest guards and communities. Second, the arena of intervention, including communities, must be rendered technical. This is challenging because communities do not exist as imagined, so project discourses make "critical omissions and erasures" that must be confronted in practice. Third, knowledge must be authorized; in CFM, the dilemma of critique versus advocacy must be contained for social scientists, which ends up in the reassertion of the authorizing will to govern and improve. Fourth, failures and contradictions are presented as resulting from things that can be corrected. For example, interventions can be top-down—as long as they are eventually taken over by the community. Fifth, politics must be contained without infringing on participants' freedom. For example, recognizing deficiencies in communities enables state forest departments to retain their role of determining policies that govern relations between communities and forests. Sixth, the assemblage must be reassembled as the ground inevitably shifts; in CFM, neoliberalism shifted the discourse to one of land tenure for market efficiency and facilitating commerce. (Li 2007:265, 268–86)

Li's assemblages are not coherent or integrated. CFM has some obvious tensions and "lines of fracture." In particular, it represents a will to improve or a will to govern but rests on the argument that communities can govern themselves (Li 2007:267). In addition, the villagers who make up the "community" often resist, largely due to a history of coercion by the forest department, which creates lines of fracture between governmentality and coercion (sovereignty) as well as villagers and foresters (Li 2007:269). And advocates for CFM like Marcus Colchester make "critical omissions and erasures" in their use of the term "community," and work hard to align their advocacy with expert intervention (Li 2007:271–2). Lines of fracture provide a way to conceptualize the contradictions in an assemblage, which are managed by compromises, but remain points of fragility.

Assemblages are not coherent or integrated because they have no author. The agents doing the work of assembling are multiple, and work without "a master-mind or a totalizing plan"; they have "no singular rationality and no essence" (Li 2007:265). In Li's view, CFM, for example, "emerged" from the struggle between villagers and forest bureaucracies over the forest, out of work done by forestry departments, conservationists, agroforestry experts, legal advocates, activists, donor agencies, and forest villagers (Li 2007:267–8). "Fuzziness, adjustment and compromise are critical to holding assemblages together," Li (2007:279) writes, because they are characterized by tensions and fractures. These tensions and fractures are what make government difficult to accomplish. Li sees government—and it is governmentality—actually being accomplished through the practices of holding assemblages together (Li 2007:287).

Mitchell's networks/metrologies can be compared to Li's assemblages. While Li sees the difficult work of assembling and holding together an assemblage as government, Mitchell emphasizes rivalry between metrological regimes. If we add the two together, we get rival assemblages, each of which is trying to conduct something, but which are also difficult to assemble and keep together.

Timothy Mitchell[3] (2008) does not use the term assemblage; instead he uses the terms "socio-technical arrangements" or "socio-technical complexes," and then, after introducing Latour, "metrologies." Socio-technical complexes, in Edison's case, included "arrangements involving patents, legal powers, political connections, capital flows, generating equipment, and power cables"; they were not "the more limited human networks of economic sociology" (Mitchell 2008:1117, 1118). Metrology, as I said above but quoting Mitchell, is "the gigantic enterprise to make of the outside a world inside which facts … can survive" (Mitchell 2008:1119, quoting Latour 1987:251). Mitchell (2008:1117) also borrows Andrew Barry's (2002) term "metrological regimes": "extensive but often fragile zones of measurement that have become relatively standardized." These arrangements or complexes or metrologies are Latour's networks; they are also assemblages.

Mitchell gives us two examples of such assemblages. His first example is Edison's Electric Light Company, which promoted the supply of electric lighting and held patents on all the devices his team invented. It had rivals, especially the Westinghouse Electric Company. It assembled "socio-technical complexes" that tied together "humans and electrons, the flow of electric current and the flow of capital, imagination and illumination, the calculation of the cost of copper wiring and of its conductivity," "connecting generators, light bulbs, buildings, shoppers, consumer desire, and capital investment" (Mitchell 2008:1117). All of the things in these complexes "helped to constitute the world that would gradually take shape and be identified as 'the economy'" (Mitchell 2008:1119).

Mitchell's second example is the economic "discovery" that poor people (in Peru) who gained title to their property (through the Peruvian development economist Hernando de Soto's project) began to work harder, which seemed to confirm the tenets of neoliberal economics (Mitchell 2008:1119). After briefly debunking this finding,[4] Mitchell (2008:1120) traces the history of the production of this "fact," through de Soto and his connection to the Chicago school's neoliberal movement, to his position as chief adviser to the Fujimori government, arguing that he was able to "install the metrology that made the later findings possible." Mitchell (2008:1120) argues that "economic facts were established in a world that was organized, through specific projects such as de Soto's property titling program, to enable economic knowledge to be made."

Economic calculations are an essential part of both examples. Calculations were "caught up in" Edison's complexes (Mitchell 2008:1118). De Soto's project helps Mitchell argue that the economy itself is "caught up in" these projects (Mitchell 2008:1119). The success of economic calculations is determined by whether or not they help create a metrological regime. A calculation is successful, Mitchell writes, if it makes it possible "to conceive of a network, or market, or national economy, or whatever is being designed, and assist in the practical work of bringing it into being." Economic calculations helped "to bring into being the world they calculated"; accuracy was less important than whether or not the calculations "enabled the network to be conceived and built" (Mitchell 2008:1118). Finally, calculations are conceived against rival calculations, and must "persuade others that they are superior to rival models and calculations" (Mitchell 2008:1118). Economic calculations are practices of assemblage.

Mitchell (2008:1116) examines the practices and work that answer the question of "how" a particular object, the modern economy, is produced. He argues that the economy, in its modern sense, emerged in the mid-twentieth century as a reified object of calculation, as something that always already exists, as a pre-existing sphere—but it is actually an invention built out of rival projects, like Edison's electrical system and de Soto's neoliberalism, to establish metrological regimes. The economy is a "product" or "invention" of networks or metrologies: "Metrologies create and stabilize objects; the economy is a very large instance of such an object" (Mitchell 2008:1119). This means that assemblages can create and stabilize objects too. CFM might be seen as another such object, rather than simply an example.

Another product of socio-technical complexes or metrologies, Mitchell argues, is the organizing of distinctions, of binaries. For example, the social and the economic, the material and the cultural, cannot be separated in any systematic way in these complexes, but they have been seen in these terms (Mitchell 2008:1117). The virtual/real binary is similarly part of the networks. For example, Edison's networks are simultaneously virtual and real: "Every instance of building [electrical] networks was simultaneously a demonstration and the thing being demonstrated, something virtual and something real" (Mitchell 2008:1119). The production of binaries is part of the work that creates an object like the economy. I argue that the distinction between assemblages as projects and assemblages of effects is another such boundary.

Assemblages of effects

Mitchell (2002) traces the effects of the raising of the Aswan dam's height in Egypt in 1933, a British colonial project, revealing heterogeneous elements and heterogeneous linkages between them. "War, disease, and agriculture" formed a web connecting "rivers, dams, fertilizers, food webs" and more, through linkages which were "hydraulic, chemical, military, political, etiological, and mechanical" (Mitchell 2002:27). "Dams, blood-borne parasites, synthetic chemicals, mechanized war, and man-made famine coincided and interacted" (Mitchell 2002:22). Mitchell notes that:

> there are no accounts that take seriously how these elements interact. It is as if the elements are somehow incommensurable. They seem to involve very different forces, agents, elements, spatial scales, and temporalities. They shape one another, yet their heterogeneity offers a resistance to explanation.
>
> *(Mitchell 2002:27)*

I am arguing that these sorts of clearly interrelated but incommensurable effects of interventions also constitute a sort of assemblage. Mitchell (2002:52, 53) uses terms like "heterogeneity," "the mixed way things happen," "inseparable mixture," and "impossible multiplicity." He raises two questions about these heterogeneities: "How exactly did tanks and parasites and synthetic nitrates affect one another?" and "What kind of explanation can bring them together?" or how can we explain them (Mitchell 2002:22)? Both are "how" questions.

From the perspective of an environmental anthropologist, the most interesting things about Mitchell's heterogeneities is that they include the non-human and that Mitchell analyses how the significance of the non-human was excised from history. His main goal in this piece is to provide an explanation. His first argument is that:

> no explanation grounded in the universalizing force of human projects and intentions can explore whether the very possibility of the human, of intentionality, of abstraction depends on, at the same time as it overlooks, nonhuman elements.
>
> *(Mitchell 2002:29)*

Social theory, he says, sees "nature, tools, obstacles, resources" as essentially passive—and such a view cannot begin to think about this heterogeneity (Mitchell 2002:29). This assemblage gives an active role to nature.

Mitchell's second argument is that in spite of this heterogeneity, "forms of rationality, planning, expertise, and profit arise" (Mitchell 2002:30). Out of this interconnected but unanticipated mess, in other words, people imagined a world "that somehow seems the outcome of human rationality and programming" (Mitchell 2002:30). The dam, to take one element of this mix, came to be seen as a triumph of science, human ingenuity, blueprints, and ideas over passive nonhuman elements. Through an argument against the idea of personified capital, Mitchell provides the first suggestion of an answer:

> The answer I want to propose here has to do with the role of expertise and reason, explanation and simplification, in the politics of the twentieth century. Politics itself was working to simplify the world, attempting to gain for itself the powers of expertise by resolving it into simple forces and oppositions.
>
> *(Mitchell 2002:34)*

The main simplifications and oppositions involved were "nature versus science, material reality versus human ingenuity, stonework versus blueprints, objects versus ideas" (Mitchell 2002:36).[5]

The assemblages described by Mitchell were imagined, "misapprehended" discursively, in order to construct expertise, capitalism, and a nation-state—in other words, to construct power:

> This misapprehension was necessary, for it was exactly how the production of techno-power proceeded. Overlooking the mixed way things happen, indeed producing the effect of neatly separate realms of reason and the real world, ideas and their objects, the human and the nonhuman, was how power was coming to work in Egypt, and in the twentieth century in general.
>
> *(Mitchell 2002:52)*

Actual, real failure does not disrupt this power. Failures and adjustments are overlooked or "actively covered up" (Mitchell 2002:42). New projects "were themselves responses (and unsuccessful responses) to problems caused by earlier technoscientific projects, in particular the Aswan Dam" (Mitchell 2002:42). The power that Mitchell sees coming to work in the twentieth century is the power to ignore heterogeneity by viewing the non-human as passive, in order to construct things like plans and expertise and capitalism.

Mitchell ends by addressing the question of what we can do. He suggests making "power and agency a question, instead of an answer known in advance," and acknowledging "the unresolvable tension, the inseparable mixture, the impossible multiplicity, out of which intention and expertise must emerge" (Mitchell 2002:53).

My addition to this is essentially that the concept of assemblage might allow us to think about interconnections between the human and non-human, the material and the discursive, in the effects of interventions. Adding assemblages of effects to purposeful assemblages brings in lots of elements of interest to environmental anthropology in general; this seems particularly true of the anthropology of conservation, in which the interactions of things like existing ecologies and economies with expert plans is so often surprising.

Conclusion

There is power in all three readings for today, but it is seen quite differently in each piece. In Mitchell 2008, the emphasis is on how rival projects create a world, a metrological regime, to accommodate their ideas. Metrologies, also called socio-technical complexes, are assemblages. The real power in this piece is the power to make a virtual idea into a real thing that survives in the world. This is what assemblages do. Li (2007) sees governmentality as difficult to achieve, and highlights the tensions and fractures (many of them political) in a governmental project like CFM, and the hard work of adjustment and compromise that holds this assemblage together. Assemblage works for Li as a way of thinking about how governmentality accomplishes government. The power in Mitchell (2002) is primarily discursive, stemming not from the accomplishment of an assemblage, but from the way it is misapprehended. The projects he describes are, like Li's, also governmental. But the power he ends up describing is much larger: it is the power to make things like technoscientific expertise or capitalism seem real.

We can add these pieces together. Assemblages may typify governmentality, whether in CFM or irrigation in Egypt. It does take work to assemble them. These governmental assemblages may be part of much larger assemblages of effects, often responded to with new governmental assemblages. Assemblages may conjure relatively stable "objects" like CFM, the electric industry, the economy, expertise, capitalism. These objects are also discourses, which may depend on binaries, like those dividing the human from the non-human and the instrumental from the discursive, that require forgetting a lot about the assemblages that built them.

We need to be cautious about the role that we assign to human actors in constructed assemblages. As Li (2007:286) says of her piece: "I emphasized agency—the work that situated individuals do—without re-inscribing the self-sovereign subject with a master-mind, a master-plan or a singular interest and intention." The assemblage described in Foucault's quote in the introduction of this chapter does not even include people.[6] His assemblage/apparatus does have a "strategic function" in response to a need, e.g., the imperative of a floating population became the matrix that undertook the control of madness, both in terms of new disciplines of knowledge and of new institutions (Foucault 1980:195). The apparatus is thus "always inscribed in a play of power" (Foucault 1980:196). But strategies do not emerge from a singular or self-sovereign author.

Foucault also notes that as soon as an assemblage/apparatus is born, a "double process" of strategies and effects begins. First, "each effect—positive or negative, intentional or unintentional—enters into resonance or contradiction with the others and thereby calls for a re-adjustment or a re-working of the heterogeneous that surface at various points" (Foucault 1980:195).

Second, there is "a perpetual process of *strategic elaboration*" in which unforeseen effects are "re-utilized for diverse political and economic ends" (Foucault 1980:195–6, author's italics). What Li and I have called practices of assemblage, which produce assemblages born with strategic functions (perhaps in Mitchell's competitions of power), and effects of assemblages are interrelated processes. Li's idea of lines of fracture in assemblages also works with Foucault's effects in contradiction with others. Every assemblage as intervention spins off an assemblage of effects, which can spin off another assemblage as intervention, etc. In fact, the two assemblages are not really separate.

Assemblages of whatever kind are hard to think with, but the concept promises a way to analyze that brings us closer to ethnographic questions about "how" things happen. And perhaps to thinking about nature and society together.

Notes

1. Note their "how" question.
2. Tania Li is a professor of anthropology at the University of Toronto in Canada.
3. Mitchell is a British-American political theorist and historian and professor of Middle Eastern studies at Columbia University. He is married to the anthropologist Lila Abu-Lughod.
4. Mitchell 2005 documents this case study in full detail.
5. Mitchell (2002:312n77) differentiates his argument and James C. Scott's.
6. I owe this observation to Javier Gonzalez Rivero.

References

Arnold, Ross D. and Jon P. Wade 2015 A Definition of Systems Thinking: A Systems Approach. *Procedia Computer Science* 44:669–78.

Barry, Andrew 2002 The Anti-Political Economy. *Economy and Society* 31(2):268–84.

Deleuze, Gilles and Félix Guattari 1987 *A Thousand Plateaus: Capitalism and Schizophrenia*. Minneapolis, MN: University of Minnesota Press.

Foucault, Michel 1980 *Power/Knowledge: Selected Interviews and Other Writings 1972–1977*. Colin Gordon, ed. New York: Pantheon Books.

Latour, Bruno 1987 *Science in Action: How to Follow Scientists and Engineers through Society*. Cambridge, MA: Harvard University Press.

Li, Tania M. 2007 Practices of Assemblage and Community Forest Management. *Economy and Society* 36(2): 263–93.

Mitchell, Timothy 2002 Can the Mosquito Speak? In *Rule of Experts: Egypt, Techno-Politics, Modernity*. Pp. 19–53, 209–43. Berkeley, CA: University of California Press.

Mitchell, Timothy 2005 The Work of Economics: How a Discipline Makes Its World. *European Journal of Sociology*. 46(2):297–320.

Mitchell, Timothy 2008 Rethinking Economy. *Geoforum* 39:1116–21.

16

UNIVERSALS, COLLABORATIONS, AND GLOBAL AGREEMENTS

Introduction

In this chapter I am going to explore Anna Tsing's (2005) ideas about universals and collaborations. I will compare this to Annelise Riles' (1998) analysis of the patterned process of creating a UN agreement. Both universals and global agreements are essentially assemblages.[1] Both have the power to summon the global. But, as we will see, both are also vacuous, meaningless.

This chapter is highly relevant to conservation discourses, collaborations, and global agreements. The main universal Tsing explores is capital "N" Nature, and her examples are thus all environmental. Riles' UN agreement involves gender, but international environmental agreements are negotiated in very much the same way. It is our universal that has power and is vacuous. It is this powerful, empty universal around which our collaborations gel. It is our international agreements that bridge differences with empty language.

For Tsing, universals raise the question: how does the universality of Nature operate in a world of cultural differences? How does the universal, "Nature," forge connections between people with very different ideas about what nature is? For Riles, global agreements raise the questions: How do global agreements achieve their effects, and what effects do they achieve? How do they work, and what work do they do? These are all "how" questions. And I ask, how on earth does the lack of meaning create connections and agreements?

Tsing and Riles share many of the same theoretical references, including Foucault and Latour. In addition, they both did ethnographic field research on collaborations, in Indonesia and Melanesia, respectively.

Universals

Anna Tsing[2] (2005:89) asks: "How does the universality of Nature operate in a world of friction?" Friction, the title of her book, is defined as "the grip of worldly encounter" or "the vicissitudes"—conquest, competitive passions, resistance strategies, confusions, struggles, etc.—concealed by a teleology like "progress" (Tsing 2005:1,6). It is productive or creative (Tsing 2005:ix–x). She writes:

> Friction makes global connections powerful and effective. Meanwhile, without even trying, friction gets in the way of the smooth operation of global power. Difference can disrupt, causing everyday malfunctions as well as unexpected cataclysms. Friction refuses the lie that global power operates as a well-oiled machine. Furthermore, difference sometimes inspires insurrection. Friction can be the fly in the elephant's nose.
>
> *(Tsing 2005:6)*[3]

The word "difference" or "cultural difference" is often used as a synonym for "friction." In this world of friction, universals become not truths but "sticky engagements" (Tsing 2005:6, 267). To re-word her question: how does the universality of Nature work in a world characterized not by truth, teleology, or global power, but by the stickiness of real differences and the vicissitudes of real histories of engagement between differences? Tsing says that to be effective, universals must engage. As she writes in her conclusion:

> This book has argued that we know and use nature through engaged universals. The "environment" spreads around the world through the friction of engagement, both for commercial users, who tap into its divergences for capitalist commodity chains, and for advocates, who find in these same divergences the means to study, enjoy, or preserve it.
>
> *(Tsing 2005:270)*

Tsing answers her question by considering the process of generalization to universal. Generalization, she says, is the product of the interaction of "axioms of unity," which establish compatibility or commensurability among disparate facts (the "fruit" that unites apples and oranges), and "tentative and contingent" collaborations or convergences among "disparate knowledge seekers and their disparate forms of knowledge," which bridge difference (Tsing 2005:89–90). "Axioms of unity and collaborations both need each other and hide each other," she writes: "The specificity of collaborations is erased by pre-established unity; the a priori status of unity is denied by turning to its instantiation in collaborations" (Tsing 2005:89–90). Together, axioms and collaborations produce the generalizations that make claims to global scale, to universality, possible (Tsing 2005:89–90).

In Foucault's terms, universals are discourses that have been elevated to the status of truths. Truths participate in Foucault's *dispositif* assemblages, but not people (see Chapter 15). What Tsing does is add social relations to truth; in fact, she gives social relations in the form of collaborations a role in producing universals. Tsing's universals can be seen as assemblages more like Latour's than Foucault's. However, she adds the interactions between discursive axioms and social collaborations, which "need each other and hide each other." And her key is bridging differences: axioms bridge differences between different facts, and collaborations bridge differences between knowledges and knowledge seekers. Tania Li is certainly aware of things that are hidden and the need to bridge difference in practices of assemblage; for example, the erasures that occur when social reality doesn't fit the discourse, forging alignments with compromises across lines of fracture, containing dilemmas, and reframing contradictions. Li's analysis is more detailed; Tsing's more sweeping. But Tsing's (2005:112) analysis of universals is a thoughtful introduction to more ethnographic chapters, which "track globe-making interactions much closer to the ground," as we will see in the following section.

Tsing gives us four examples of universal Nature: botanical science, which used God's will and then evolution as axioms of unity and drew on historical collaborations between botanical collectors;[4] American nature loving, which used a spiritual, aesthetic unity for Nature and drew on collaborations with tourism promoters; climate modeling, which used systems logic and desired collaborations between scientists and policy-makers; and sustainable forest management, which used international politics and failed to create effective collaborations.

I am going to briefly review two of these examples, botany and climate modeling. Botany was the first global science—Tsing says it paved the way for the global scale and the universal in science—but it was historically based on erased collaborations between European and non-European knowledge and collectors, as we saw in Chapter 8. Botany began by interpreting God's work as a system of classification, which meant that every new plant discovered in fifteenth and sixteenth century voyages of exploration had a place in a pre-existing logic (Tsing 2005:90–1). Eventually this a priori system produced the theory of evolution, which allowed the logic of nature itself to usurp the logic of God as the axiom of unity. The actual, locally specific facts about plants (and many of the categories of the system of classification) depended on collaborations with non-Europeans. These came to be completely hidden: "European botanists came increasingly to imagine themselves as communing directly with plants—and the universality of science—without the mediation of non-European knowledge" (Tsing 2005:91). Political economy is also relevant here: the growing political power of Europe contributed to this imagining, constituting an "imperial gaze" (Tsing 2005:94–5). Tsing ends this example by reminding us that botany is "the touchstone for biodiversity discourse" and "a resource for environmental politics" (Tsing 2005:95). To sum up, the first science to claim the global, universal scale for science, a science that continues to contribute to global environmental politics, began at the juncture of the assumption that God's work was logical with local knowledge of plants, and continued by shifting God's logic to Nature itself and erasing local knowledge entirely.

The second example I am going to detail is contemporary global climate modeling, in which science, rather than being established as a new universal, essentially becomes the "a priori" unifying axiom: "The model declares itself continually to be science. It is expert, neutral, rational, and empirically grounded" (Tsing 2005:102). Collaborations, rather than providing the local knowledge on which the model is based, are the end goal of climate modeling: "Neither botanists nor nature lovers have used globality to push potential collaborators to the negotiating table, but climate modelers do just that" (Tsing 2005:102). Modelers want to bring in policy-makers. As Tsing (2005:103) argues: "It is my contention that the global commitments of the model are strategic: They are tuned to stimulate international dialogue." They also, she continues, cover up dialogue. In order to do so, they use the global scale to supersede "nations, classes, cultures, or specific business interests" (Tsing 2005:103). The local disappears inside the global (Tsing 2005:104). She uses an ethnography of a 1995 conference on the human dimensions of global modeling to present three "surprises." First, "The global scale takes precedence—because it is the scale of the model"; second, "Models breed more models"; and third "Models must be charismatic and pedagogical" (Tsing 2005:103–5). These three aspects of global climate modeling, she suggests, did produce negotiations (e.g., Rio 1992 and Kyoto 1997), but they did not produce consensus. Instead, two primary conflicts emerged: North versus South and United States versus Europe. She ends with the result of these conflicts: the US, which combined a difference in chosen universals ("private property, free trade, and the entrepreneurial spirit") with unilateral power (note a political-economic factor, again), withdrew from the Kyoto Protocol in 2001. In sum, global climate modeling depends on science as its axiom of unity and has collaborations with policy-makers as its only partially achieved goal; the local disappears, as do less-powerful voices.

Tsing is arguing that there is nothing natural about the global scale, that Global Nature is not naturally a universal; universal Nature and the global are constructed through the "hard work" of projects like botany and climate modeling. The construction of Global Nature rests on things that are not natural: assumptions of God's logic, the theory of evolution, the circular logic of a global scale taking precedence because it is the scale of the model. Global Nature "facilitates and obscures worldwide collaborations" (Tsing 2005:111).

Universals that can essentially create the global scale, that can facilitate global collaborations and obscure them too, are powerful discourses. When Tsing (2005:265) turns to more ethnographic material, she calls them "charismatic" universals, which have "power to travel" but are also "easily co-opted."

Collaborations

Collaboration for Tsing, like compromise for Li, is a double-edged sword—and the double edges are similar. Li said that compromise is an agreement between two parties who each have some sense of betraying themselves (see Chapter 14). Tsing (2005:245) begins her chapter with the dictionary definition of collaboration, as "united labor, cooperation" or "traitorous cooperation with the enemy." Neither of them is writing about anything like consensus.[5]

In the final chapter of *Friction*, Tsing retells a collaboration she wrote about in 1999.[6] As she reveals in the book's preface, this collaboration played an important role in her imagining of the book (Tsing 2005:x). This collaboration stimulated and exemplifies Tsing's idea of friction. In this final chapter, a universal—"Nature"—momentarily coalesces a collaboration, which is remembered quite differently by each collaborator; the collaborators also define "Nature" quite differently. Nevertheless, the collaboration succeeds in carving a real community forest out of state plans involving timber concessions. This is a success story, but not a simple one.

Tsing (2005:246) says that collaborations have "friction" at their heart. Friction, as we have seen, is what cultural differences bring to global connections. Cultural differences cause misunderstandings, which are productive and positive in Tsing, because misunderstandings allow people who are different to connect. Misunderstandings are essential to global connections, and they are at the heart of collaborations. Misunderstandings are creative; they make new objects—in this case community-managed forests—and new agents possible (Tsing 2005:247). Misunderstandings depend on the "power to travel" of charismatic universals, like Nature, community empowerment, and forest conservation. Discursive fields forge connections seemingly based on understandings, which conceal underlying misunderstandings. Charismatic universals are powerful discourses. Tsing, however, is not emphasizing their power, but *how* they work. She is not emphasizing "abstract principles of power and knowledge," but "concrete engagements," for she is an ethnographer, "a listener and a teller of tales" (Tsing 2005:267–71).

None of the collaborators in this ethnography agree about what actually happened. First, environmentalists in Banjarmasin (also known as Kompas Borneo), the provincial capital of South Kalimantan, Indonesia, are double experts who both revere and want to develop the Dayak people. They see Nature as a wild place, and understand "community" as a cultural unit held together by traditional rituals. In their eyes, the Dayak in the village of Manggur are on the threshold of modern history, primitive and longing to be modern (with their expert help). Second, organizers in the national capital of Jakarta use the example of Manggur to help build a policy model in which Nature is populated (not wild) and managed by rational, resource planning "communities." In their view, the Dayak are not primitive or modern, but engaged in a continuous struggle to manage their own resources. They don't, however, really care about the Dayak at all. Third, leaders in the village of Manggur copy the rhetoric of external authority, hoping to hold together their community with the legitimacy they gain by doing so. For most Dayak, Nature plays an essential role in social group formation. For Manggur leaders (the same sibling set showcased in Tsing 1999), the community forest became the medium for networking with external authority, and stories about how it was won are a claim about their personal community leadership abilities.

Tsing (2005:268) ends the book with hope located in the provincial activists:

> All of us need a little of the romanticism and commitment of these urban middle-class young people. Through them, the utopianism of a mobilization for the environment as justice becomes a challenge rather than a reason to give up.

She calls for researchers to ground their analyses of power and knowledge not in abstracts but in local places where "universals," hopeful discourses, are situated or engaged; this, she promises, should give us "critical purchase" "without cutting off the springs of hope" (Tsing 2005:267). She says that activists spend a lot of time trying to create consensus or arguing over differences, a theme we will return to in the following section (Tsing 2005:262). She would like them to allow misunderstandings and differences.

Tsing makes it clear that universals are not meaning-full and collaborations not built on consensus; on the contrary, "Nature" is empty, a black hole that absorbs differences while engaging collaborators, who can connect in spite of those differences.

Global agreements

In her analysis of the fascinating process of producing global agreements, the legal anthropologist Annelise Riles[7] (1998) argues that the process of creating UN agreement documents in global conferences, which she compares to the piling up Fijian mats (Riles' research was among Fijian delegates), creates consensus among "levels" (local, national, regional, global)—not through meaning, but through an exercise of form. She argues that the UN problem of bringing together levels in a "global" agreement is not solved through a meaningful discussion of political differences, but rather through dictates of form, of standardized structure, of aesthetic pattern. She allows us to add this exercise of form to Tsing's engaged universals. We might also add it to Li's practices of assemblage (see Chapter 15).

Two years of preparation for a UN conference included local and then national and regional conferences and meetings, producing local and then national and regional document drafts, for a final, global-level conference: the 1995 UN Fourth World Conference on Women, which produced and ratified the Platform for Action and the Beijing Declaration (Riles 1998:378–9). Succeeding conferences ascend increasing levels of generality. In the three-week process of the final meeting, documents are created (and shared with all delegates) that combine lower-level drafts, marking differences between them by putting them within brackets (which can also contain multiple levels of brackets). Every paragraph bears a digital set of numbers and letters that mark its place in the series that is the document; documents also bear these identifiers. A running count is kept of the brackets. The object of the final meeting is to agree on language, suggested by delegates, that bridges differences within and thus eliminates brackets, and is often literally quotations (without using quote marks) from previous "clean" (bracket-free) global documents. Negotiators at the conference were not interested in the meaning of this language. Riles thought that "the string of words in the document consisted of a jumble of words and paragraphs that seemingly pointed nowhere" (Riles 1998:388). She asks: "How does this institutional knowledge [UN agreements] achieve its effects, and what effects does it achieve?" (Riles 1998:388).

I argue that this is another way, comparable to Tsing's engaged universals, to obscure differences and create the global. Some of Riles' language is the same as Tsing's, though the two do not cite each other. In Tsing, axioms of unity and bridging collaborations together produce the generalizations that make it possible to claim global scale. In Riles, odd, standardized, aesthetic conventions direct delegates to bridge local differences until they achieve a global agreement document.

Parallels between the uses of UN documents and Fijian mats, especially the grouping of mats into "*vivivi*" sets, allow Riles to explore pattern and scale in the aesthetics of negotiation. The aim of producing a mat or a document was to replicate a pattern given at the start or make a good specimen of a fixed genre (Riles 1998:381, 386). Both documents and mats were "collective, anonymous, and highly labor-intensive exercises that required great attention to detail"; both exercises "yielded objects collectively acknowledged as highly valuable"; like mats, documents "provided the concrete form in which collectivities … were 'taken to' another environment"; finally, both were also "items of collection … elements of sets" (Riles 1998:382).

The "aesthetic device" of both *vivivi* and document is the alternation between pattern/abstraction and object/concreteness (Riles 1998:388). In a ceremonial space, mats are layered such that the boundaries between them are "no longer foregrounded"; they are only "pattern" (Riles 1998:384–5). Once the ceremony has ended, they are tossed into piles by type and counted; they have become objects (Riles 1998:384–5). A document is a pattern until it is "clean," when it becomes an object. Documents, like mats, follow a "simple nonrepresentational patterning":

> The character of the pattern—a simple logic that linked words, paragraphs, documents, or conferences—entailed the collection of a potentially infinite number of concrete and distinct entities (words, paragraphs, conferences) into a straightforward digital sequence of numbers and letters.
>
> *(Riles 1998:385)*

The pattern for these documents, like the *vivivi*, is "given at the start" (Riles 1998:388), and documents produced at each level of negotiation—local, national, regional, international—mirrored the same pattern. The objective of this patterning was not meaning, but rather aesthetics (Riles 1998:386): "One had to acquire an ear and an eye for the patterns here" (Riles 1998:387). The pattern is evident only in the process of producing a document, which is a kind of "sorting exercise" comparable to the layering of mats (Riles 1998:387). Once the "clean" document is produced, delegates take it home and take it apart again, reshuffling the text into material for quotation in later documents (Riles 1998:393).

Brackets allow a view of infinity. Within the brackets is "a black hole of meaning" that encompasses the potential infinity of layers of language from layers of other documents produced at lower levels (Riles 1998:390–1). A running count is kept of the number of brackets and the percentage of the document within brackets, which Riles argues only makes the "infinity" within them more evident. Negotiators mask that "infinity within the brackets" by quotation: language that "held firm and did not reveal other layers" (Riles 1998:393). This infinity within the brackets is an infinity of differences and politics.

Quotation provides concreteness. It bridges the gaps in the brackets, eventually eliminating the brackets altogether. The meaning of a quotation was taken as being "self-evident"; what mattered was its concreteness, as it moved from conference to conference and document to document. This concreteness constituted the concreteness of the document as "clean" object. "The concrete document was an encapsulation of the infinity of pattern within something that stood on its own, just as the *vivivi* could be rolled up and taken away" (Riles 1998:393). The clean document is the global imagined as a concrete object, as "real." I argue that quotation is self-evident because it has a little bit of the quality of universals. The "clean" text is the only consensus here.

Finally, these aesthetic devices, bracketing and quotation, bring into view the local and global: "What the patterns in the language of the *Platform for Action and the Beijing Declaration* ultimately represented, what we might understand as its meaning, was precisely the 'levels' (global, national regional, and so on) that the designers of UN procedures so desperately sought to bring together" (Riles 1998:394). In other words, the UN document achieves the effect of a consensus of levels by emphasizing meaningless patterns until an "object" emerges.

Conclusion

Universals and UN agreements conjure the "global." Universals are generalizations that make claims to global scale; generalizations combine axioms of pre-established unity and specific collaborations, which both need each other and hide each other. UN agreements do this by bringing lower-level drafts together through a process that ultimately erases them in order to produce a "clean," concrete document—which is woven out of quotations from past "clean" documents that bridged differences. The process through which UN agreements conjure the global is collaborative, though the collaboration works to erase lower-level differences and draw bridging language from past documents. Over the course of a few years, the collaboration also works to generalize, to ascend levels from local to national to regional to global. There is also a similarity between the language of past documents and generalizations: both are empty of meaning.

Universals are empty, truths based on a priori assumptions of truth. That is actually the source of their power to support collaborations, which are based on unexplored misunderstandings. UN agreements are also empty, cleansed of the potentially infinite differences inside the brackets. Global consensus depends on erasing differences. Meaning actually depends on difference. Meaning does not conjure the global.

In some ways the differences that are erased by universals and UN agreements are local. The lack of acknowledgment of local sources of botanical knowledge is like the insertion of lower levels of documents into brackets in agreement documents. But Riles suggests that the local, as well as the global, is conjured by the UN process. She says: "The local and global are brought into view through the aesthetic devices described in the previous pages"; "the very contexts or perspectives of global and local are artifacts of the practices we wish to understand" (Riles 1998:394).

Tsing is doing something quite different, reminding us that universals *are* local even though they may erase this fact: her universals are situated or engaged in the local places where they are used. This allows her to call for the utopianism of a mobilization for the environment as justice. It gives ethnographers critical purchase without cutting off hope, allowing us to watch vacuous universals like Nature being productive and creative. It gives activists reason for optimism—though they do have to allow misunderstandings and differences. I think it gives conservationists reason for hope too.

But what about global environmental agreements? I think the question is: other than being picked apart to provide bridging language for future agreements, how are they used? How are they situated or engaged in local places?

Notes

1. Li's assemblage included both "authorizing knowledge," which might be compared to Tsing's universals, and "forging alignments," which creates collaborations.
2. Anna Lowenhaupt Tsing is a professor in the Anthropology Department at the University of California at Santa Cruz.
3. This quote is reminiscent of Tsing 2009 about the role of differences in supply chain capitalism, covered in Chapter 12.
4. As we saw in Richard Grove's history, in Chapter 8. Tsing draws on Grove's work for her analysis of botany.
5. Tsing's ethnography here concerns community-based conservation and community-based natural resource management, almost directly parallel to Tania Li's ethnography of community forest management, which she considered in terms of practicing assemblage (see Chapter 15).
6. I have written about Tsing 1999, in Dove and Carpenter 2008.

7. Annelise Riles is Jack G. Clarke Professor of Law in Far East Legal Studies and a professor of anthropology at Cornell University. Her work focuses on the transnational dimensions of law, markets, and culture. She has conducted legal and anthropological research in China, Japan, and the Pacific.

References

Dove, Michael R. and Carol Carpenter, eds. 2008 *Environmental Anthropology: A Historical Reader*. Malden, MA: Blackwell.

Riles, Annelise 1998 Infinity within the Brackets. *American Ethnologist* 25(3):378–98.

Tsing, Anna L. 1999 Becoming a Tribal Elder, and other Green Development Fantasies. In *Transforming the Indonesian Uplands*. Tania M. Li, ed. Pp. 159–202. Netherlands: Harwood Academic Publishers.

Tsing, Anna L. 2005. *Friction: An Ethnography of Global Connection*. Princeton, NJ: Princeton University Press.

Tsing, Anna L. 2009 Supply Chains and the Human Condition. *Rethinking Marxism* 21(2): 148–76.

17

WORLD-MAKING IN THE ANTHROPOCENE

Introduction

The Anthropocene is a term in common and informal scientific usage for the current geological age, when humans have dominated the climate and the environment. Tsing considers that it begins with the advent of capitalism, and many others see it this way. It is a term that unfortunately enshrines the degradation story,[1] and the opposition between people (anthropo-) and the earth assumed by that story. What if the story were changed? This chapter explores an emerging literature that uses natural history to tell stories of assemblages that include humans and non-humans making worlds in collaboration: world-making. Such stories may help us rethink survival in our current age.

My intention in this book has been to give the reader a tool-box of ways to think about power in conservation. The first tool was the power of discourses to affect our thinking. I contend that we need to think differently about conservation, and about human–nature relations in general. In this chapter I want to explore the work of writers that seem to me to point a way forward, to beckon us towards a new sort of ethnography that has its nose very close to the ground, and also allows us to think about the Anthropocene in new ways. This literature is relatively mute on power, but I present it as a way to begin to shake off some of the discursive power of the degradation story and the nature–culture opposition.

I begin with the trailblazer, Donna Haraway. Haraway (2003:16) writes about the "implosion" of nature and culture ("natureculture") in the historically specific, connected lives of dogs and people bonded in "significant otherness": "companion species." Anna Tsing (2014) follows, exploring the sociality of living but non-human beings, especially relations between fungi and red pine in the old satoyama landscape in Japan. Tsing (2015) then adds people and explodes her view to observe and tell stories about the contemporary supply chain of that same fungi, matsutake. Finally, Andrew Mathews describes a historical assemblage of chestnut trees, livestock, and farmers in Italy.

Haraway, Tsing, and Mathews almost certainly all know each other: they all teach at the University of California at Santa Cruz, two of them in the same department. Haraway and Tsing mention each other; Mathews has cited Tsing. They have all drawn from Foucault and Latour. Haraway has a biology doctorate, and is most known for science studies. Tsing and Mathews are environmental anthropologists, and their work is based on ethnographic research.

Companion species

Companion species are historically specific, connected lives of non-humans and people bonded in "significant otherness." Donna Haraway's[2] (2003:15) *The Companion Species Manifesto* is about people and dogs, but she mentions others, including people and rice, bees, tulips, and intestinal flora. Haraway (2003:32, 61) argues that we need to take "companion species" seriously, in thought and action, in order to "get us out of inherited boxes" and in order to learn how to care about different worlds. Companion species share significant otherness: "otherness-in-connection," "otherness-in-relation" (Haraway 2003:20, 32, 45, 49). And they implode conceptual oppositions; companion species *are* "naturecultures." This otherness-in-relation is situated (historically specific) and also partial, because companion species are irreducibly different; they are different species. Communication across irreducible difference is what matters, she says, emphasizing respect for that difference. (Haraway 2003:8, 16, 25, 49) Haraway (2003:50) says that the key is recognizing that one cannot know the other or the self, "but must ask in respect for all of time who and what are emerging in relationship."

Looking into the past, Haraway prefers to say that human and dog co-constituted each other rather than that man domesticated dog, and she likes the phrase "co-evolution," urging us to take it as the rule, not the exception. And she will not allow the dog to evolve only biologically while people evolve culturally: human genomes may contain a molecular record of dogs' pathogens (Haraway 2003:30–1). She coins the term "metaplasm," for the physical remodeling of dog and human in the history of their relations, including their DNA (Haraway 2003:20). Dogs and people are fundamentally different but not separate.

There is an ethics and a politics here; it is a manifesto. Haraway (2003:61) argues that alertness to otherness-in-relation—the alertness of dog to owner and owner to dog during training, for example, especially for agility trials—can make us "more worldly," "more alert to the demands of significant otherness at all the scales that making more livable worlds demands."

> The task is to become coherent enough in an incoherent world to engage in a joint dance of being that breeds respect and response in the

> flesh, in the run, on the course. And then to remember how to live like that at every scale, with all the partners.
>
> *(Haraway 2003:62)*

Haraway is saying that we should use the significant relations with different species to learn how to make more livable worlds together. We can learn to attend to, respect, and care about different species. And we can use these understandings to "get us out of inherited boxes," to resist discourses, like nature versus culture.

"More-than-human sociality"

Anna Tsing (2014) argues that if anthropologists are to think well about environmental change and landscape transformation, we must begin the process of writing "critical description" about more-than-human sociality. "More-than-human" embraces both the human and the non-human; "social" means "made in entangling relations with significant others" (Tsing 2014:27). "Critical description" combines the methods of ethnography with natural history; it is critical because "it asks urgent questions" (Tsing 2014:28). Anthropologists who want to understand environmental change "need to know about the social worlds other species helped to build" (Tsing 2014:33).

Tsing considers several questions, making important points along the way. First, we need to rethink historical concepts of freedom and being, which have blocked our ability to consider sociality as more than intention and planning; she suggests freedom to act rather than a willed freedom, and world-making activities rather than conscious being (Tsing 2014:29–31). Second, she argues that assemblages and forms can be used to do ethnographies of living beings who do not talk. Assemblages here are things like the plants that grow near each other in a landscape. Forms of plants like trees may show the history of sociality that has shaped them, e.g., a "pasture tree" with thick lower branches (Tsing thanks Andrew Mathews for this insight). She cautions that non-humans have autonomy from human designs, e.g., bacteria. Third, looking at human involvement with non-humans is a place to begin, but agency cannot be limited to humans: humans are "one of many historical agents" (Tsing 2014:34).

> All the varied trajectories that have made an impact on the landscape would be relevant, human and otherwise. Together these would make up the landscape's polyrhythms, that is, its enactment of multiple conjoined histories.
>
> *(Tsing 2014:34)*

In a present characterized by environmental change, more-than-human social lives are building landscapes and constructing histories together, and the methods of natural history and ethnography can be used to appreciate them.

Tsing applies this to a particular landscape, the Japanese satoyama forest, a peasant woodland characterized by deciduous oaks and red pines now in decline but being restored. The satoyama forest is shaped for human uses (wood, charcoal, mountain vegetables, chestnuts, bamboo shoots, mushrooms, forage, and green manure). It is shaped by human investments; satoyama woodlands are working landscapes in which human disturbance has shaped the forest, although not always by human design. These investments include coppicing, burning, logging, shifting cultivation, and the removal of organic matter from the forest floor for green manure. (Tsing 2014:35–6) Tsing does not see human investments as creating the satoyama's web of social relations, which emerge, rather, from relations among satoyama species. For example, red pine thrives not because it is useful or the target of any investments, but because it needs the disturbed open woodlands, the bare soils, that peasant investments create. The design of satoyama woodlands is an unintended multispecies design, in which all the species make worlds for themselves and each other. Multiple histories characterize the forest (Tsing 2014:36).

Matsutake mushrooms are an associate of red pine in satoyama woodlands. They are ectomycorrhizal fungi, wrapping around tree roots, taking carbohydrates from them and helping the trees get water and other nutrients (Tsing 2014:38). Japanese people consider them to be delicious. Contingencies, however, as always, intrude. War and peasant migration to cities, among other things, leaves the satoyama woodlands neglected, and the species mix changes to evergreen oaks and laurels. Red pine and mushrooms disappear together, with effects for humans and non-humans. Restoration is attempted, but must be indirect: matsutake mushrooms cannot be made to grow by direct human efforts. Fungi "draw us into worlds of many interacting species," a good place to start thinking about humans as participants in more-than-human sociality (Tsing 2014:39).

Anna Tsing's[3] 2015 book, *The Mushroom at the End of the World*, continues the story of matsutake from Oregon, where the mushroom's decline in Japan and a variety of other contingencies (logged and burned forests taken over by lodgepole pine) opened the Japanese market to Oregon's matsutake, now primarily an associate of lodgepole. The book tells the story of the entire supply chain of the matsutake mushroom, from the logged-over forests of Oregon to gourmets in Japan. Matsutake are one of the most valuable mushrooms in the world. Matsutake foraging is now one way to make a precarious living, without wages or benefits, in the Pacific Northwest of the US. The people who choose this way of life are often Southeast Asian refugees, motivated by freedom.

In Tsing 2014 she mentioned natural history and ethnography as the methods appropriate to a critical description. In *Mushroom* she goes into more detail. Curiosity and imagination, "the arts of noticing," Tsing argues, should constitute our methods. Telling stories rather than counting can be used to pass on what we have learned. She sets the arts of noticing and telling stories against progress stories (or stories of ruin and decay, which is just the opposite of progress) and the idea of the self-contained (non-social) individual. Progress is a "crippling assumption," as is the assumption of the self-contained individual actor maximizing personal interests (Tsing 2015:5). Unfortunately, the former has framed the study of economy and ecology, and the latter neoclassical economics and population genetics, which has made economy and ecology "algorithms of progress-as-expansion" (Tsing 2015:5, 28). To give up these assumptions requires noticing, especially noticing as in natural history and ethnography (Tsing 2015:28). "Why don't we use these stories [of contaminated diversity] in how we know the world?" Tsing asks. Because, she answers, they are "recalcitrant to the kind of 'summing up' that has become the hallmark of modern knowledge" (Tsing 2015:33). "If a rush of troubled stories is the best way to tell about contaminated diversity, then it's time to make that rush part of our knowledge practices" (Tsing 2015:34).

"More-than-Human Sociality" (Tsing 2014) conveyed a vision of more-than-human social lives building landscapes and constructing histories together; *Mushroom* embroiders this vision by using the language of assemblage (see Chapter 15). Tsing tells us that this book "sketches open-ended assemblages of entangled ways of life, as these coalesce in coordination across many kinds of temporal rhythms" (Tsing 2015:viii). Her assemblages are interspecies, open-ended gatherings, ephemeral or shifting, consisting of "entangled lifeways" (ways of being) contaminating each other productively, creating new lifeways. Her assemblages are polyphonic, autonomous but intertwined melodies (Tsing 2015:23). Spatially, they are typified by mosaics or patchiness. Time is particularly polyphonic, consisting of multidirectional histories, non-progressing temporal patterns, multiple time-making projects. There are multiple world-making projects, human and not human, in her assemblages (Tsing 2015:21). For example, preindustrial livelihoods and new ones like mushroom picking make worlds too, but are invisible because they are not part of progress (Tsing 2015:22). "All organisms make ecological living places, altering earth, air, and water" and changing everyone's world. Bacteria, for example, made our oxygen atmosphere, and fungi made soil (Tsing 2015:22).

In Chapter 16 I examined Tsing's writing on human collaborations based on misunderstandings; in *Mushroom* she expands that concept to all species, and makes collaboration the key to survival in our times. Tsing 2015 is really about survival at the end of the world; its subtitle is "On the Possibility of Life in Capitalist Ruins." Precarity is Tsing's term for life at the end; she sees it as the condition of our time. Precarity is not only life in dire times: it is the condition of being vulnerable to transformation by others, of being in relations with others; it is frightening, but makes life possible (Tsing 2015:20). "This book argues that staying alive—for every species—requires livable collaborations. Collaboration means working across difference, which leads to contamination. Without collaborations, we all die" (Tsing 2015:28). Encounters across difference contaminate us, change who we are, change our world-making projects—and this is the stuff of survival (Tsing 2015:27–8). We are changed in collaborations. New mutual worlds may emerge in such encounters: contamination makes diversity (Tsing 2015:29). Collaborations create change. "Transformation through collaboration, ugly and otherwise, is the human condition" (Tsing 2015:31). "Neither tales of progress nor of ruin tell us how to think about collaborative survival," Tsing says, but mushroom picking might open our imaginations (Tsing 2015:19). In *Mushroom* Tsing is linking collaboration to change, and thus to survival.

In Chapter 13 I presented Tsing's ideas about supply chain capitalism; in *Mushroom* she uses the matsutake supply chain to rethink how natural resources are drawn into capitalism. The matsutake chain:

> illuminates something important about capitalism today: Amassing wealth is possible without rationalizing labor and raw materials. Instead, it requires acts of translation across varied social and political spaces.... Translation ... is the drawing of one world-making project into another.... Translations across sites of difference *are* capitalism: they make it possible for investors to accumulate wealth.
>
> *(Tsing 2015:62)*

In their supply chain, mushrooms are "translated" from trophies of freedom to capitalist assets to Japanese gifts. She also terms this translation salvage accumulation.[4] "Salvaging" is coopting "living things made within ecological processes." Salvage accumulation is "the process through which lead firms amass capital without controlling the conditions under which commodities are produced" (Tsing 2015:62–3). Salvaging sites are both inside and outside capitalism; they translate value between noncapitalist and capitalist value systems (Tsing 2015:63). And "we need an ethnographic eye to see the economic diversity through which accumulation is possible" (Tsing 2015:66).

Agro-sylvo-pastoral assemblages

In Monti Pisani, central Italy, between about 1000 and 1800, cultivated chestnut forests were part of complex agro-sylvo-pastoral assemblages. These assemblages also included upland sheep and goat grazing and lowland farmers, who were dependent on upland manure for fertilizer. Premodern Italian states were linked. So were oak and pine, mycorrhizal fungi, edible *Boletus* mushrooms, bacteria, stone wall-edged terraces to retain water, and transformed soils. Andrew Mathews[5] (2017:G151, G152), like Tsing, sees this as an "assemblage."[6]

Chestnut forests were cultivated by grafting desired varieties onto wild root stock and pruning shoots below the graft. Chestnuts were the staple food in this system (Mathews 2017:G146, G151). Capitalism introduced exotic fungal diseases of the chestnut in the 1850s and 1950s, shifting the use of chestnut from food staple to fuelwood or tannin production. In this new system, chestnuts were logged and the trunk coppiced as it regrew (for tannin and fuelwood production). Today, Mathews tells us, the forests and the words for them are "ghosts."

Mathews argues that these "ghosts" are still discernable, and describes and illustrates a method much like that advocated by Tsing: the close reading of material forms through an "alert practice" of natural history, alongside the study of fading linguistic terms. The close reading of material forms in these ghost forests is enabled by the fact that chestnut trees change form. The grown trees can be "read" as either the grafted/pruned form (*selva*) of the chestnut tree or the timbered/coppiced form (*ceduo*) (Mathews 2017:G148–51). Old stumps can be read as grafted and pruned then coppiced. As chestnut forests are abandoned, the linguistic terms are lost, but *selva* and *ceduo* can be found, "embedded" in forestry regulations (Mathews 2017:G152). Material forms and linguistic terms co-emerge (Mathews 2017:G154).

Mathews describes his natural history method: he walks a transect, noting tree, shrub, and understory plant species; looking at bark for evidence of disease, grafting, or fire; looking at walls, ditches, houses, and ruins. "This is hard work," he says; "it requires constant attention to form, texture, and color, constant speculation as to pattern." He speculates as he walks: "This is mentally exhausting work that requires close attention, and yet, paradoxically, it also contains an element of speculation." It reminds him of participant observation, a key method in ethnographic research, and the notes he takes are also similar (Mathews 2017:G147).

Mathews emphasizes that he is not describing relations between pre-given entities, but rather "attending to the multiple forms that emerge from partial relations between different plants, animals, and people" (Mathews 2017:G153). A chestnut is not one thing: in its *selva* form it is in an assemblage of partial relations with goats, people, sheep, and terraces; in its *ceduo* form it is in a dense forest of pole-sized stems repeatedly cut and regrowing.

Mathews agrees with social science critiques of the term "Anthropocene" as having produced a "singular and impoverished language of politics."[7] Reading ghost landscapes in Italy, stories of chestnut diseases, and histories of economic change, he says, does not lead to this Anthropocene. He argues that Anthropocenes are "irreversibly multiple." So, for example, 1,500 years of sustainable peasant chestnut cultivation was an Anthropocene. So was the abandonment of this assemblage as disease and industrialization emerged. The ghosts of past Anthropocenes matter as "resources for contemporary environmental politics," and for "producing different visions of livable futures" (Mathews 2017:G153):

> Paying close attention to the ghostly forms of past histories in present-day forests allows us to consider the many forms of political and economic life that these forests are or might be connected to, including imagining multiple possible Anthropocene futures.
>
> *(Mathews 2017:G154)*

Conclusion

Haraway, Tsing, and Mathews give us visions with which to combat the discursive power of ideas embedded in the term Anthropocene. They also detail methods for carrying out research on that vision that add something new to ethnography. Finally, we can discern two sorts of politics in all of their work: the politics of resisting a regime of thinking about humans and non-humans that they see as harmful, and the politics of why it matters how we think.

All of these authors correct our perspective of life with other species by sharing their own visions. Haraway's vision is of a kind of relation—companion species, significant otherness, naturecultures—that teaches us how to respect and care about different worlds. She is sure that in the history of dog and human we remodeled each other. Tsing envisions open-ended assemblages of interwoven, interspecies world-making projects. She makes the social character of world-making explicit and important: we are all entangled in relations with significant others of other species, contaminated by each other in the productive sense of creating new possibilities together. Her examples are the historical matsutake–red pine collaboration in Japanese satoyama woodlands, and the contemporary matsutake supply chain. Mathews' example is also historical, and he emphasizes this: the chestnut groves of Italy are ghost forests, nearly gone but still discernible in the contemporary landscape.

All of these authors say something about their methods, and their methods emphasize something like alertness or attention coupled with curiosity or speculation. Haraway repeatedly emphasizes the alertness required to communicate with the significant other. Her example is the human–dog relation during agility trials. Tsing emphasizes the arts of noticing, based on curiosity and imagination, and combining natural history and ethnography. She also thinks about writing, calling for telling stories (e.g., of collaborative survival) and critical description. Mathews writes about an "alert practice" of natural history—close attention coupled with speculation—alongside the study of fading linguistic terms.

Haraway, Tsing, and Mathews are all thinking about survival, and how they can contribute to thinking about how to survive. Haraway thinks that learning how to respect and care for different worlds though companion species relations may be the key to learning how to make more livable worlds. In "More-than-Human Sociality" (2014) Tsing writes generally about environmental change and landscape transformation, but in *Mushroom* (2015) the ruins are the ruins of capitalism; she is thinking about how we will all survive the ruins of capitalism together. Her key is collaboration. Her example is matsutake harvesting, a lifeway for pines, mushrooms, and people, even while it is also on the peripheries of capitalism. She addresses the relation between capitalism and these assemblages: in salvage accumulation, living things made in ecological processes are translated into capitalist value. For Mathews, the ghosts of old forests matter because they can produce different visions of livable futures.

Haraway thinks that being in a companion species relation, and thinking about it, can help us get "out of inherited boxes," and Tsing and Mathews agree. Tsing talks about the many boxes she thinks we need to get out of: freedom defined by will, being defined by consciousness, "counting," progress stories, and self-containment. These assumptions, she says, cripple our thinking. Mathews faults the impoverished language of the Anthropocene, arguing that we need to be able to imagine multiple Anthropocene futures.

Their visions all give power in the sense of agency or autonomy to the non-human, which the idea of the Anthropocene does not do. Haraway and Tsing insist that dogs and fungi have agency and autonomy from humans, and have things we do not usually associate with non-humans: dogs have moments of loving communication, fungi have social relations with trees, and both have their own world-making projects. Tsing argues that if we think of freedom in terms of action rather than intent, and of being as world-making rather than consciousness, we open ourselves to seeing the agency and autonomy of non-humans—which we must do to think about landscape transformation. In Mathews the autonomy of the chestnut emerges in its form, which is multiple, emerging from its ability to change form in relation with the humans and non-humans around it.

These realizations seem to me to be essential to thinking about conservation. Insofar as conservation follows the same logic as the Anthropocene—that people are destroying non-human species—conservation must separate people from those endangered non-human species. But if the visions shared by these authors are illuminating, then conservation is wrongheadedly wresting apart assemblages of conjoined world-making projects. If these authors are right, then survival is going to happen through these social relations and assemblages. We need to learn to see them, and to think them.

Notes

1. See Chapter 3. The term "degradation story" comes from Fairhead and Leach, and is a term for the common assumption, not always true, that people degrade the environment.
2. Donna Haraway is Distinguished American Professor Emerita in the History of Consciousness Department at Santa Cruz, and has a Yale PhD in biology. Haraway is a biologist specializing in science studies, who uses Latour and Foucault, among others, and mentions Tsing in this piece.
3. Anna Tsing is a professor in the Anthropology Department at Santa Cruz, and has a Yale BA; her PhD is from Stanford. Tsing mentions Latour and Haraway, among others. Tsing's 2014 piece has the recognizable form of an academic article, but her 2015 is much more eclectic. As she says, the book's form and thesis follow each other (Tsing 2015:viii). Tsing et al. 2017, which Mathews appears in, is also unusual, consisting of two parts, "Ghosts" and "Monsters," one printed front to back and one back to front.
4. *Friction*, using the language of frontiers, is also essentially about salvage accumulation.
5. Andrew S. Mathews is an associate professor in the Department of Anthropology at the University of California, Santa Cruz. He has a 2004 PhD in environmental anthropology (joint degree School of Forestry and Environmental Studies/Anthropology), Yale

University; a 2000 MPhil in anthropology, Yale University; and a 1995 MSc, *Forestry and Its Relation to Land Use*, Oxford University, England. In different publications, Mathews has used Foucault and Latour, and much of the literature in this book, including Tsing. He is the author of *Instituting Nature* (2011), a history of conservation and forest management in Mexico. He is "currently working on a book on the historical ecology, natural history, and climate politics of Italian forests, very tentatively titled *Plant Politics*": "Because Mediterranean ecosystems have evolved to cope with dramatically variable climate, powerful disturbances, and intense human modification they are good places to learn about climate change and the Anthropocene" (https://amathews.sites.ucsc.edu/about-me/, accessed 11/26/2018).
6. Mathews uses both the term system and the term assemblage.
7. See for example Lövbrand et al. 2015.

References

Haraway, Donna 2003 *The Companion Species Manifesto: Dogs, People, and Significant Otherness*. Chicago, IL: Prickly Paradigm Press.

Lövbrand, Eva, Silke Beck, Jason Chilver, Tim Sorsythe, Johan Hedren, Mike Hulme, Rolf Lidskog, and Eleftheria Vasileiadou 2015 Who Speaks for the Future of the Earth? How Critical Social Science Can Extend the Conversation on the Anthropocene. *Global Environmental Change* 32(April):211–18.

Mathews, Andrew S. 2011 *Instituting Nature: Authority, Expertise, and Power in Mexican Forests*. Cambridge, MA: The MIT Press.

Mathews, Andrew S. 2017 Ghostly Forms and Forest Histories. In *Arts of Living on a Damaged Planet: Ghosts*. Anna L. Tsing, Heather Swanson, Elaine Gan, and Nils Bubandt, eds. Pp. G145–56. Minneapolis: University of Minnesota Press.

Tsing, Anna L. 2005. *Friction: An Ethnography of Global Connection*. Princeton, NJ: Princeton University Press.

Tsing, Anna L. 2014 More-than-Human Sociality: A Call for Critical Description. In *Anthropology and Nature*. K. Hastrup, ed. Pp. 27–42. London: Routledge.

Tsing, Anna L. 2015 *Mushroom at the End of the World: On the Possibility of Life in Capitalist Ruins*. Princeton, NJ: Princeton University Press.

Tsing, Anna L., Heather Swanson, Elaine Gan, and Nils Bubandt, eds. 2017 *Arts of Living on a Damaged Planet:* Minneapolis: University of Minnesota Press.

18

CONCLUSION

Introduction

I begin by briefly reviewing the theoretical tools in this book and their relevance for conservation. I then return to the issue of Marx and Foucault raised in Chapter 1. Marx and Foucault take me to "how" questions. Here I link them to ethnography, and ethnography to relatively invisible domains. I examine linkages between power and the economy, and the effect of discursive power on our ability to "see" those linkages. A short note on ecology in the book follows. I end by considering how we might not govern in conservation.

A tool-box of ideas about power in conservation

Foucault's power in general[1]

We should think in terms of power relations that are multiple and mobile, not capital "P" Power. Power always involves free subjects. Though both parties in a power relation may have intentions, the larger-scale strategy that may coalesce out of these intentions in combination with others' intentions is not the result of an individual's intention. Domination may gel for a time, but is always accompanied by resistance.

Any intervention, for conservation or something else, is an act of power, and involves implementers in power relations with local peoples. They are free subjects, who may resist.

The power of discourses[2]

Foucault sees power in the knowledge that is accepted as truth; knowledges come to be considered truth in a political process (the intentions of power relations), and considering something to be true privileges it with power. We are typically not conscious of these truth regimes; for the most part, people simply believe that what is defined as truth *is* truth.

"Truth" thus has power in Foucault, but its power isn't what we're used to thinking about as power. No one forces us to believe in a truth. Discursive power cannot be wielded by institutions, not even by states. Nor can discourses be deliberately controlled, not even by the seemingly powerful: institutions do intend to create discourses (e.g., using public relations or advertising science), but they cannot make them catch on—they cannot give them power. Finally, a single "truth" can serve resistance as well as institutions with political-economic power. In Foucault's terms, we cannot "speak truth to power."

Discourses accepted as true have the power to shape what we think and do, even (especially) in science and policy. People are not the source of discourses, even those that they themselves believe, and speak or write. But Foucault's method offers us a resistance: if we analyze "truth" historically—or ethnographically—we can come to question even what we ourselves believe.

Conservation is a discourse with power in Foucault's sense, in a field of linked and powerful discourses: nature, wilderness, the environment, sustainability, biodiversity, endangered species, etc. How did this rich discursive field come to gather around and join in the huge, complex project of conservation interventions in the global South? And how has this discursive field and its power influenced the effectiveness of conservation in the real world? These questions cannot be answered without a full and specific understanding of what discourses and their power are.

Knowledges with truth-power also have power over knowledges that lack the power of truth, as we saw in Grove's history and Li and Nadasdy's ethnographies.[3] This power, the power of one knowledge over another, also has an important, negative, impact on conservation. It means that we cannot incorporate "local knowledge" of the environment into scientific environmental conservation without disemboweling it. It makes it very difficult to listen to or learn from those with other knowledges.

Sovereignty, discipline, governmentality, and the triangle[4]

Foucault differentiates between three models of government that we can separate analytically but which all occur together in reality, including in conservation interventions.

The biggest difference is between sovereignty on the one hand and discipline and governmentality on the other: sovereignty does not govern life; discipline and governmentality do. I have argued that discipline should not be confused with governmentality. The two together govern life, not only at different scales, but in very distinct ways. Disciplinary power dreams of a perfect structure, in which everything and everyone can be ordered, which leads to regulations and surveillance, and a huge administrative state. Governmentality thinks quite differently, of improvement and optimization, rather than perfection. It tries not to govern too much. But mechanisms of security are necessary to manage and plan for the freedoms of the population and the economy.

When conservation is implemented, people and the environment are governed. Each type of government has different characteristics and implications for the effectiveness of conservation. The relation between types of government is of particular importance.

Sovereign government usually occurs in park conservation, because parks are territorial units often governed by excluding certain people and restricting uses. These exclusions and restrictions on use are sovereignty, particularly when violence or the threat of violence is used to enforce them. Even a past history of displacement, exclusion, and violence has consequences for efforts to govern behavior outside the park.

Discipline is still often used to govern people's behavior in and around parks or in areas where wildlife is being protected. Surveillance, multiple regulations, and enforcement constitute discipline. Anything that resembles the military, prisons, factories, or schools may be disciplinary. Conservation has a tendency towards disciplinary government because it is based on the degradation discourse, which imagines local people as a threat to the environment that must be stopped. This "norm" of conservation and the discipline that it engenders do not sit well with more liberal models of government.

Governmentality is the most liberal model of government. Its rationality is a positive one: improving people's welfare. But the intent is still to change behavior, to "conduct conduct." In conservation the intent is to make people conserve. The organizations doing conservation act on actions by managing and planning; this is governmentality. It determines our stance towards the environment as well as the people using it. People are difficult to manage and perceptive about the power it implies.

The triangle of government clearly exists in conservation. Ethnographic examples in Africa and Indonesia saw histories of displacement and exclusion from parks, coupled with violence; together with discipline and surveillance of buffer zone populations; together with participatory approaches attempting to cultivate conserving subjects with trainings and contracts. In one case detailed by Li, the combination led to park invasion.

Conservation's knowledges, biology and ecology, and governmentality's knowledges, management and economics, have the power to prevent conservationists from seeing the power and government inherent in conservation—which local people can see.

States, centers, and blinding magic[5]

Scott's power is simplification itself and the way simplification creates legibility to a state, which uses it to reshape nature and society. Early modern states believed that they could and should improve nature and society, using science. Scott's insights about power are applicable to conservation. Conservation organizations exist to improve nature and society using science. Like high-modernist states, they displace people and their traditional uses of nature and attempt to replace them with something considered more conserving in both the scientific and economic senses. They do not appreciate how local society limits their ability to do this.

Latour's ideas about the accumulation and simplification of immutable mobiles and metrology adds a layer to the understanding of how scientists and bureaucrats in conservation build influence and work to extend it into the world.

Tsing explicitly expands the sorts of knowledge that inform simplifying interventions to include the "neutral" knowledge through which the environment is "explored, managed, converted, and conserved," which depend on calculative practices of efficiency, natural capacity, and profit margins. She brings conservation into theories of simplifying. Power in Tsing is a magic that blinds ecologists and economists (and conservationists) to the fact that their own approach to nature is also cultural. More importantly, it blinds them to local natural history: a nature filled with the cultural and the social, and which might provide new and better tools to think about forest conservation in areas of human settlement.

Subject formation[6]

Subjects are constructed by knowledges that study people and use them as units of analysis, especially the human sciences. They are cultivated by disciplinary mechanisms like surveillance, pressure to conform to norms, the maintaining of files and cases on individuals, and by architecture that separates them and manipulates them to observe themselves. Finally, subjects are created by the very concept of individuality, by encouraging people to know and improve themselves.

Many conservation projects working with people intend to create conserving subjects. Agrawal's ethnography of forest councils in India examines a success and analyzes how it occurred: participation in practices of government. Li's ethnography of integrated conservation and development projects in Indonesia, in contrast, details failures. Subject formation is difficult, and can be risky.

Neoliberal governmentality[7]

Whereas governmentality governs the population, neoliberal governmentality governs subjects. It does so in two ways, by improving their acquired human capital, thus ensuring the innovations that allow for economic growth, and by manipulating the milieu in which economic subjects calculate the costs and benefits of their conduct. Only the latter is used in conservation.

Conservation occurs in a world defined by neoliberal governmentality. It seems natural to conservation policy-makers and implementers to see the people with whom they cross paths as economic subjects. It is easy for us to believe that we can address conservation problems by introducing alternative sources of income and stepping up the costs of degrading. We act on the economic milieu of people we imagine as economic subjects in order to govern their conduct. We need to become more aware of what we are doing. Tania Li would argue that the hill people in Sulawesi are making calculations, and that they know that the benefits we have to offer don't outweigh what they're giving up in the park (see Chapter 10). Li also notes that REDD+ projects are asking farmers to accept money for not using forests, but a lot less money than timber is worth. Underlying these two sorts of conservation projects is a contradiction between market sensibilities, which price natural resources, and conservation, which needs people to accept much less than market worth. We need to be less naïve about this contradiction.

Invisibility of implementation; assemblages; universals, collaborations, and agreements; and world-making

Chapters 14 through 17 of this book introduce four ideas that are important to conservation: the invisibility of project implementation; practices of assemblage and assemblages of effects; universals, collaborations, and global agreements; and world-making.

Implementation[8] is the site where governmental plans—whether by policy-making organizations or states—meet reality. It is the "witches' brew" of practices that Foucault did not study, but ethnography does. Real conservation, the conservation that is written on real landscapes, happens here. But this space is invisible. Its invisibility is built into the structure of conservation interventions as a space that doesn't have to be directly ruled. Most of the realities of implementation do not reach policy or state bureaucracies, whose politics depend on interpretations. State rule is accomplished, and policy interventions "succeed," in interpretation. Whether we characterize this as governmentality or not, it seems to me to describe much of conservation. It may explain why there is so little evaluation of conservation interventions. But it is deeply troubling.

Assemblages[9] include people and ecologies; in fact, they may be useful to think about environment–human relations. Conservation cannot be accomplished without practices of assemblage, as in Li's example of community forest management. Assemblages of effects accompany all projects, certainly including conservation. Assemblages may be a useful way to think about what we are really doing. They include discourses, and discourses may work on them. In the case of conservation, practices of assemblage always include the environmental resource to be conserved. Assemblages of effects probably always include the non-human as well, though development projects may excise these elements retrospectively. Mitchell, for example, analyses how the significance of the non-human was excised from the history of the development of Egypt.

Universals, collaborations, and global agreements are all highly relevant to conservation.[10] The main universal Tsing explores is capital "N" Nature, and international environmental agreements are negotiated in the same way as Riles' UN agreement. It is the conservation universal that has power and is vacuous. It is this powerful, empty universal around which conservation collaborations gel. It is international conservation agreements that bridge differences with empty language.

Haraway, Tsing, and Mathews share their own visions of world-making in order to correct our perspective on life with other species.[11] Haraway's vision is of companion species, significant others, or "naturecultures" that can teach us how to respect and care about different worlds. Tsing envisions open-ended assemblages of interwoven, interspecies world-making projects. Her world-making is explicitly social: we are all entangled in relations with significant others of other species, in the productive sense of creating new possibilities together. Her examples are the historical matsutake–red pine collaboration in Japanese satoyama woodlands, and the contemporary matsutake supply chain. Mathews' example is the ghost chestnut groves of Italy, nearly gone but still discernible in the contemporary landscape.

Their visions all give power to the non-human, which the idea of the Anthropocene does not do. If we think of freedom as action rather than intent, and of being as world-making rather than consciousness, we can see the agency and autonomy of non-humans. Conservation follows the same logic as the Anthropocene, that people are destroying non-human species, a logic that separates people as the only world-making agents from nature as victim. I fear that conservation in the Anthropocene is wrongheadedly wresting apart assemblages of conjoined world-making projects.

Marx and Foucault

I argued in Chapter 1 that Marx-based and Foucault-based approaches to understanding power in conservation are both essential—but that we cannot continue to over-simplify Marx and Foucault's ideas. Political ecology tends to over-simplify Marx, and when Foucault's ideas are embraced they tend to be over-simplified too. I argue that Escobar uses Marx and Foucault together in a way that over-simplifies the power of discourses (Chapter 3).

But this book provides examples of ways in which Marx and Foucault may be used together. Braun's history of capitalist extraction "tangled together" with governmentality in the response of the Canadian state to the new science of geology is a good example (Chapter 10). Comparing Li's *Will to Improve* and *Land's End* demonstrates the appropriateness of the theories to different sites. Li also uses different, specific concepts from Marx to analyze each site (Chapter 13). Tsing's commodity chain capitalism (Chapter 10) and salvage capitalism (Chapter 17) demonstrate an ethnographic eye for the economic probably based on Marx, but going where the ethnography takes her. All these practices are exemplary: considering how capitalism and governmentality are tangled together, using different theories for different research sites, using specific concepts from Marx (or Foucault), and following ethnography away from theory.

I worry that over-simplified ideas of power vilify conservation by linking it unfairly to structural forces of capital "P" Power. We need to listen to and learn from conservationists if they are to listen to and learn from us. I think it is more productive to assume that conservationists are well meaning, though some of them need to learn more about power.

"How" questions, ethnography, and invisible domains

As we have seen, Foucault asks how, not why, questions, and so do several of the authors in this book. To ask "why" about power—why is power exercised in conservation?—is to ask a question that is not researchable. It only makes sense if we think we already know the answer. It sends our thinking to some pre-determined source of power. It does not question the singularity of Power. To ask "how" questions about power—How is power exercised in conservation? How is power accomplished in conservation?—is researchable. As Agrawal (2005:210, my italics) says of political ecology: "The primacy accorded the political often prevents political-ecological analysis from examining *how* the political itself is made." It "casts politics as the prime mover, the cause that exists sui generis. But the exercise of power and political asymmetries … are themselves a consequence of many different processes" (Agrawal 2005:210). "How" questions are intimately linked to the ethnographic method. As Mosse (2004:641, author's italics) says, "My concern is not whether, but *how* development works. The approach is *ethnographic*." Li (2007:264) says that "the answers to 'how' questions … deserve a more central place in studies of government than is currently accorded." These questions concern the *practice* of government, and can only be answered ethnographically. Tsing (2005:267–71) notes that she is not emphasizing "abstract principles of power and knowledge," but "concrete engagements," for she is an ethnographer.

"How" questions that have been asked in this book include Foucault's (1983:216–17) own: H ow is power exercised? Agrawal (2005:224–5) asks: How do problems that require government come into being?[12] How do new subjectivities form?[13] Mosse (2004:641) asks: How does development work?[14] How is policy success produced?[15] Li (1999:296) asks: How is government or rule accomplished?[16] How is an assemblage assembled (Li 2007:254)? How is contestation manifested in an assemblage? Given their fragility, how are assemblages secured, i.e., what practices secure them?[17] Mitchell (2008:1116) asks: How is the modern economy produced? "How exactly did tanks and parasites and synthetic nitrates affect one another?" "What kind of explanation can bring them together?" (Mitchell 2002:22).[18] Tsing (2005:89) asks: How does the universality of Nature operate in a world of cultural differences? How does the universal

"Nature" forge connections between people with very different ideas about what nature is?[19] Riles (1998:388) asks: How do global agreements achieve their effects, and what effects do they achieve? How do they work, and what work do they do?[20] Whatever aspect of power in conservation is addressed, there is a "how" question for it.

Whether or not the authors examined in this book explicitly champion ethnography, they are almost all ethnographers. Foucault's own theory is based on the detailed study of historical texts. It is in many ways comparable to ethnography, except that it cannot produce understandings of ideas *in practice*—and ethnographies can. Scott incorporates an ethnographic understanding into theory about power. For example, local society, practical knowledge, and nature, as he knows from ethnographic research in Malaysia, resist complete simplification. Latour's ideas about building networks of allies and extending the laboratory into the outside world are based on his research in laboratories.[21]

Running throughout this book is an odd theme about things forgotten, devalued, simply not seen, or deliberately put aside. In Foucault (1978:100, my italics) this is part of discursive power: discursive elements are distributed between "the things said and those *concealed*, the enunciations required and those *forbidden*." Ferguson's seminal *The Anti-Politics Machine* documents the power of the discourse of development to make political contexts and effects disappear.[22] Tsing rails against the "blinding magic" of economics and ecology to make local society and culture disappear.[23] The knowledge of the local physician Achuden is forgotten in the history of botany.[24] Scott's essential mētis is devalued and ignored by states.[25] In Mitchell developers of the Aswan dam do discursive work to eliminate the importance of the non-human.[26] Li argues that integrated conservation and development knows but puts aside structural political and economic forces.[27] Ethnography is the best research method for these hidden things. They are so important to effective conservation that remembering them, valuing them, seeing them, and not putting them aside should constitute a separate tool.

Two domains are particularly invisible: implementation (Chapter 14) and household-level economies (Chapter 13). Implementation is remarkably invisible to the policy-makers and the states or organizations behind them. Mosse argues that the practices of implementation are in a black box. I suggested that there is an anti-implementation machine parallel to Ferguson's anti-politics machine. This invisibility is not ignorance; project planning and reporting processes simply exclude the complex and political realities of project implementation. Ethnography like Mosse carried out is the best method for studying project implementation.

Household-level economies are invisible to the discipline of economics, which dominates development (and increasingly conservation). These economies combine eating and selling: corn to eat is grown with onions to sell in Colombia; wheat to eat and sugarcane to sell are grown by fellow villagers in Egypt. And eating and selling are not kept apart: Colombian farmers sell corn to buy fertilizer for onions; Egyptian wheat farmers labor for sugarcane farmers in exchange for fodder and loans to get them through to the next food harvest. Ethnography is also the best method for studying these complex, and shifting, household economies. Only ethnography can study household economies meeting economic reform. For example, Mitchell was able to document the return to self-provisioning in Egypt because he carried out ethnographic fieldwork. IMF and USAID experts, in contrast, saw national statistics showing that wheat production had increased as proof that their free market reforms had succeeded.[28]

Ethnography may also provide a way to escape discourses that have blinding magic. Mosse argues that "ethnographic writing opens a rift between different epistemologies"; "to reflect and to write means striving to break free from … the discursive hold of even one's own cherished policy discourse" (Mosse 2005:x–xi). This is why Mosse champions the ethnographic approach.

Tsing and Mathews both combine ethnography with natural history, and this expansion of methods is particularly appropriate for environmental anthropologists (Chapter 17). Tsing's "arts of noticing," curiosity and imagination, are characteristic of both ethnography and natural history. Mathews argues for the close reading of material forms through an "alert practice" of natural history combined with speculation.

To sum up, "how" questions can prevent the rush to previously known answers about power in conservation, and ethnography is the best method for pursuing them. Ethnography also allows the researcher to explore invisible things, and two invisible domains important to understanding conservation: project implementation and household economies. Ethnography can help us escape the power of policy and other discourses to limit our thinking. Natural history combines nicely with ethnography.

Economy, power, and blinding magic

We need to learn to think of power and the economy together; we should not reduce the political to the economic. In Foucault's thought, government and economy are intimately related: during the disciplinary period, when the state tightly regulates commodities; during the capitalist, governmental period, when the state, informed by the new knowledge of economics, backs off from the economy; and during the neoliberal period, when neoliberals become the constant critics of state practices. Capitalism benefited from governmental techniques that made people useful; government benefitted from capitalist production, which used people. In the neoliberal period, government in general (including in conservation) acts on the "market milieu," systematically modifying its variable to shape conduct (see Chapter 9). As Mitchell notes, the neoliberal idea of free markets thrives in a world where markets are controlled by trading corporations and governmental price supports (Chapter 13). It is the "blinding magic" of economics that makes it hard to think of power and economy together.

Scott and Latour also think power and economy together. Scott (1998:8) says that both capitalism and the state simplify and render legible: capitalism and states are both agencies of "homogenization, uniformity, grids, and heroic simplification." Commercial and bureaucratic logic are "synonymous." For Latour, capitalists, economists, scientists, and engineers all share the same history of "centers of calculation," and capitalism itself may be just another process of "immutable mobilization" (Chapter 7). The blinding magic of economics makes this hard to think.

Braun and Li see intimate relations between capitalism and governmentality in ethnographies. Braun notes capitalist extraction and governmentality together at the heart of the new knowledge of geology; a newly known natural resource leads directly to extraction and government. Li argues that capitalism and improvement are "locked in an awkward embrace," because economic growth impoverishes some people, which must be seen to be managed by improvement (which failed in her case) (Chapter 10). Again, the blinding magic of economics makes it hard to think like this.

James Ferguson describes neoliberalism as a "new governmental rationality" that seeks not only to reform states but to create "new modes of subjectification," specifically new citizen and community subjects that are responsible for government results without state intervention. Structural adjustment made projects like microcredit and community-driven development necessary, as citizens and communities had to take on responsibility for their own development. If Li is right, then neoliberalism is posing itself as the solution (through community-driven development) to large-scale problems caused by (structural adjustment) or at least supported by (timber concessions) neoliberalism (Chapter 12). Neoliberalism blinds neoliberals.

We also need to think the "economy" (or "market" or "capitalism") and self-provisioning (or "subsistence") together. Gudeman and Rivera, Mitchell, and Li all provide us with ethnographies in which there is no opposition between the two. In Mitchell the opposition is only discursive, something that capitalism relies on to establish its discursive universality. Mitchell argues that local logics inflect each appearance of capitalism. In Gudeman and Rivera, peasants talk in terms of the opposition, but in reality they survive by mixing them. In Li's more isolated research site, cacao introduced a point of compulsion where those doing well had to keep buying land and planting, and those doing poorly were not left with enough land to grow food. The separation between "market" and subsistence occurred in large part because of the ecology of cacao and rainforest soils, without comprehension by the local people, and with disastrous results for many of them (Chapter 13). Economics blinds us to the intimate relations between the market and self-provisioning.

Thinking well about power and economy matters because economies *are* ecologies. In Aristotle's and Colombian peasants' terms, only the earth or the "land" is productive. Ecologies are our "base"; eating and selling both depend on ecologies (Chapter 13). Tsing's (2015:62–3) concept of "salvage accumulation" makes this explicit: "salvaging" is coopting "living things made within ecological processes." Salvaging sites are both inside and outside capitalism; they translate value between noncapitalist and capitalist value systems. Ethnography sees this, she says (Chapter 17). The economics that assumes the necessity of economic growth does not.

Learn to think about power and the economy together. Don't reduce the political to the economic. Question the existence of the market or the economy or capitalism: these are theories, not real things. Capitalism is not the same everywhere or every time. Economic reforms are inflected by people's responses to them. Local, domestic, self-provisioning, subsistence economies are not separate from capitalism. Nor are they necessarily poverty. Nor are they necessarily degrading of the environment. Stimulating economic growth will coopt more natural resources and impoverish more people; it will not solve poverty or degradation. Ecology is not separate from economy. Parks and other conservation regulations cut off options for farmers. This can cripple household economies, and may not be compensated for by integrated conservation and development or REDD+ projects. Conservation needs to see itself as asking people to give up economic options. Economics is blind to this.

Ecology

The best environmental ethnographies fully realize the influence of ecology. Donald Moore grounds theory about power in landscapes and their ecologies (Chapter 6). The ecologies of cacao and rainforest soils contribute to the way that Marx's concept of compulsion plays out in highland Sulawesi, Indonesia, as analyzed by Tania Li (Chapter 13). Timothy Mitchell describes the role of curly pondweed, *Anopheles* mosquitos, and the malaria parasite in Egypt's disasters of the 1940s, all enabled by a dam and irrigation (Chapter 15). Roy Ellen links the ecology of sago and its ambiguously wild and domesticated character to Nuaulu non-oppositional ideas about nature and culture (Chapter 4). The best environmental histories also respect ecology: Cronon says that environmental histories must respect ecological realities (Chapter 4).

Considerations of ecology as a science, however, bring power in. Paul Nadasdy, for example, contrasts First Nations peoples' understandings of ecology with ecological science, which separates species and counts animals (Chapter 8). Anna Tsing contrasts the cultural and social aspects of honey hunting in Kalimantan, Indonesia with an ecology that sees only "neutral matters of efficiency, natural capacity, and profit margins" (Tsing 2003:24) (Chapter 7).

Conclusion: ideas about how not to govern

Most policy-making and implementing conservationists must work within a system that governs people and nature, and understanding power and government may help do this less naively and more thoughtfully. But I do believe that conservation would be more effective without violence, surveillance, or the manipulations typical of "participation." As we work within this system we can also begin to imagine other ways conservation might be done. In this conclusion to my conclusion, I'm going to revisit some of the comments made by authors about how this might be done.

We may try to govern differently or cultivate subjects differently, but we cannot eliminate power. Our relations with our colleagues and our "target" populations are imbued with power. But we can learn to be acutely aware of power relations.[29] Most importantly, we can work to resist the power of discourses to constrain our thinking.

First, remember Agrawal's key to successfully creating conservation-mindedness in India's forest councils: participation in government (Chapter 10). Some of this participation was false; people imagined autonomy they didn't actually have. But in the best possible world the forest department would have given up its control of timber and resin, and in fact its legal ownership of forests. We can hold this up as an ideal: allow people to participate *in the government* of natural resources.

Second, Nadasdy offers two solutions: that traditional knowledge might be used to rethink unexamined assumptions about how people should relate to the world around them, which unconsciously form the basis of scientific wildlife management, and that Canada should give decision-making control of their territories to First Nations' peoples themselves (Chapter 8). The first point is about discursive power: conservationists might learn something from local people. The second point is about government: people might be allowed to govern their territories by themselves. This goes beyond Agrawal's participation in government, because Nadasdy's ethnography revealed that traditional knowledge could not be integrated with science; the truth-power of science molded "traditional knowledge" into its categories and counting and mapping practices. Science prevented scientists from learning anything from local people.

Finally, Tsing frames her analysis of Meratus honey hunting as offering a tool for a respectful multi-cultural conversation about tropical forest management (Chapter 7). She argues that we must improve national and international frameworks for listening. She sees relationships maintained between people, honey trees, and bees as an example of the "open-ended, semi-autonomous management relations" that make it possible for people and forests to live together for a long time, in contrast to European models, which reduce natural resources to matters of efficiency, natural capacity, and profit margins (Tsing 2003:48). The "blinding magic"—the discursive power—of science, economics, and other European models makes these respectful conversations, this listening and learning, very difficult. This book is my contribution.

Notes

1. See Chapter 1.
2. See Chapter 2 for theory, Chapters 3, 4 (conservation), 8, 13 (economics), and 17 for ethnographic uses.
3. See Chapter 8.
4. See Chapter 5 for theory, Chapter 6 for ethnographic uses.
5. See Chapter 7.

6. See Chapter 9 for theory, Chapters 10 and 12 for ethnography.
7. See Chapter 11 for theory, Chapter 12 for ethnography.
8. See Chapter 14.
9. See Chapter 15.
10. See Chapter 16.
11. See Chapter 17.
12. See Chapter 1.
13. See Chapter 10.
14. See Chapter 14.
15. See Chapter 14.
16. See Chapter 14.
17. The preceding three questions are in Chapter 15.
18. See Chapter 15.
19. See Chapter 16.
20. See Chapter 16.
21. See Chapter 7 on Scott and Latour.
22. See Chapter 3.
23. See Chapter 7.
24. See Grove in Chapter 8.
25. See Chapter 8.
26. Chapter 15.
27. Chapter 12.
28. See Chapter 13.
29. Raffles 2003 may help think about this.

References

Agrawal, Arun 2005 *Environmentality: Technologies of Government and the Making of Subjects.* Durham, NC: Duke University Press.

Foucault, Michel 1983 Afterword: The Subject and Power. In *Michel Foucault: Beyond Structuralism and Hermeneutics*. 2nd edition. Hubert L. Dreyfus and Paul Rabinow, eds. Pp. 208–26. Chicago: The University of Chicago Press.

Li, Tania M. 1999 Compromising Power: Development, Culture, and Rule in Indonesia. *Cultural Anthropology* 14(3): 295–322.

Li, Tania M. 2007 *The Will to Improve: Governmentality, Development, and the Practice of Politics.* Durham, NC: Duke University Press.

Mitchell, Timothy 2002 Can the Mosquito Speak? In *Rule of Experts: Egypt, Techno-Politics, Modernity*. Pp. 19–53, 209–43. Berkeley, CA: University of California Press.

Mitchell, Timothy 2008 Rethinking Economy. *Geoforum* 39:1116–21.

Mosse, David 2004 Is Good Policy Unimplementable? Reflections on the Ethnography of Aid Policy and Practice. *Development and Change* 35(4): 639–71.

Raffles, Hugh 2003 Intimate Knowledge. UC Santa Cruz: Center for Global, International and Regional Studies. Retrieved from https://escholarship.org/uc/item/1p69t326. Originally: 2002 Intimate Knowledge. *International Social Science Journal* 54(3)173:325–35.

Riles, Annelise 1998 Infinity within the Brackets. *American Ethnologist* 25(3):378–98.

Scott, James C. 1998 *Seeing Like a State: How Certain Schemes to Improve the Human Condition Have Failed.* New Haven: Yale University Press.

Tsing, Anna L. 2005 *Friction: An Ethnography of Global Connection.* Princeton, NJ: Princeton University Press.

INDEX

For Product Safety Concerns and Information please contact our EU
representative GPSR@taylorandfrancis.com
Taylor & Francis Verlag GmbH, Kaufingerstraße 24, 80331 München, Germany

www.ingramcontent.com/pod-product-compliance
Ingram Content Group UK Ltd.
Pitfield, Milton Keynes, MK11 3LW, UK
UKHW021721240926
475702UK00029BA/1740
9780367342500